Error-Tolerant Biochemical Sample Preparation with Microfluidic Lab-on-Chip

Microfluidic biochips have gained prominence due to their versatile applications to biochemistry and health-care domains such as point-of-care clinical diagnosis of tropical and cardiovascular diseases, cancer, diabetes, toxicity analysis, and for the mitigation of the global HIV crisis, among others. Microfluidic Lab-on-Chips (LoCs) offer a convenient platform for emulating various fluidic operations in an automated fashion. However, because of the inherent uncertainty of fluidic operations, the outcome of biochemical experiments performed on-chip can be erroneous even if the chip is tested a priori and deemed to be defect-free. *Error-Tolerant Biochemical Sample Preparation with Microfluidic Lab-on-Chip* focuses on the issues encountered in reliable sample preparation with digital microfluidic biochips (DMFBs), particularly in an error-prone environment. It presents state-of-the-art error management techniques and underlying algorithmic challenges along with their comparative discussions.

- Describes a comprehensive framework for designing a robust and error-tolerant biomedical system which will help in migrating from cumbersome medical laboratory tasks to small-sized LoC-based systems
- Presents a comparative study on current error-tolerant strategies for robust sample preparation using DMFBs and reports on efficient algorithms for error-tolerant sample dilution using these devices
- Illustrates how algorithmic engineering, cyber-physical tools, and software techniques are helpful in implementing fault-tolerance
- Covers the challenges associated with design automation for biochemical sample preparation
- Teaches how to implement biochemical protocols using software-controlled microfluidic biochips

Interdisciplinary in its coverage, this reference is written for practitioners and researchers in biochemical, biomedical, electrical, computer, and mechanical engineering, especially those involved in LoC or bio-MEMS design.

Emerging Materials and Technologies
Series Editor: Boris I. Kharissov

Atomic Force Microscopy for Energy Research
Cai Shen

Self-Healing Cementitious Materials
Technologies, Evaluation Methods, and Applications
Ghasan Fahim Huseien, Iman Faridmehr, Mohammad Hajmohammadian Baghban

Thin Film Coatings
Properties, Deposition, and Applications
Fredrick Madaraka Mwema, Tien-Chien Jen, and Lin Zhu

Biosensors
Fundamentals, Emerging Technologies, and Applications
Sibel A. Ozkan, Bengi Uslu, and Mustafa Kemal Sezgintürk

Error-Tolerant Biochemical Sample Preparation with Microfluidic Lab-on-Chip
Sudip Poddar and Bhargab B. Bhattacharya

Geopolymers as Sustainable Surface Concrete Repair Materials
Ghasan Fahim Huseien, Abdul Rahman Mohd Sam, and Mahmood Md. Tahir

Nanomaterials in Manufacturing Processes
Dhiraj Sud, Anil Kumar Singla, Munish Kumar Gupta

Advanced Materials for Wastewater Treatment and Desalination
A.F. Ismail, P.S. Goh, H. Hasbullah, and F. Aziz

Green Synthesized Iron-Based Nanomaterials
Application and Potential Risk
Piyal Mondal and Mihir Kumar Purkait

Polymer Nanocomposites in Supercapacitors
Soney C George, Sam John and Sreelakshmi Rajeevan

Polymers Electrolytes and their Composites for Energy Storage/Conversion Devices
Achchhe Lal Sharma, Anil Arya and Anurag Gaur

Hybrid Polymeric Nanocomposites from Agricultural Waste
Sefiu Adekunle Bello

Photoelectrochemical Generation of Fuels
Edited by Anirban Das, Gyandeshwar Kumar Rao and Kasinath Ojha

Emergent Micro- and Nanomaterials for Optical, Infrared, and Terahertz Applications
Edited by Song Sun, Wei Tan, and Su-Huai Wei

For more information about this series, please visit:
www.routledge.com/Emerging-Materials-and-Technologies/book-series/CRCEMT

Error-Tolerant Biochemical Sample Preparation with Microfluidic Lab-on-Chip

Sudip Poddar and Bhargab B. Bhattacharya

CRC Press is an imprint of the
Taylor & Francis Group, an **informa** business

First edition published 2023
by CRC Press
6000 Broken Sound Parkway NW, Suite 300, Boca Raton, FL 33487-2742

and by CRC Press
4 Park Square, Milton Park, Abingdon, Oxon, OX14 4RN

CRC Press is an imprint of Taylor & Francis Group, LLC

ISBN: 978-1-032-11380-7 (hbk)
ISBN: 978-1-032-11385-2 (pbk)
ISBN: 978-1-003-21965-1 (ebk)

DOI: 10.1201/9781003219651

Typeset in Nimbus Roman
by KnowledgeWorks Global Ltd.

Publisher's note: This book has been prepared from camera-ready copy provided by the authors.

Dedication

To our parents, teachers, and students

Contents

SECTION III Design Automation Methods

SECTION IV Summary

SECTION V Appendix

Foreword

During the past two decades, microfluidic lab-on-chips (LoC) have gained utmost prominence due to their versatile applications to biochemistry and medical science such as point-of-care clinical diagnosis, drug discovery, and synthetic biology. The emergence of this technology has revolutionized medicine by enabling the replacement of bulky biochemical instruments with tiny microfluidic biochips that can automate most of the necessary functions using very little sample fluid and reagents. With the help of in-built sensors, these chips can instantly diagnose a disease just from a few drops of blood or body fluid. An LoC can integrate several laboratory functions on a small chip while providing high-throughput screening and automation. A recent article (2021) published by National Heart, Lung, and Blood Institute, NIH, USA, has announced that these "tiny chips are finally starting to emerge from the lab and are poised to make an impact". Recent research on the future of LoC businesses also indicates that the consumer market for these chips is extremely promising. It was valued approximately at USD 4 Billion in 2015 and is expected to reach around USD 9 Billion by 2025.

Among different classes of LoCs, digital microfluidic biochips (DMFBs) have become very popular because of their simplicity and ease of operations. They are capable of actuating nano-/pico-liter-sized discrete droplets on a 2D-array of electrodes under electrical control. Thus, they support low-cost and fast implementation of a variety of biochemical protocols. In particular, DMFBs have received wide acceptance due to the flexibility of reconfiguring microfluidic modules while executing multiple operations concurrently on the chip so as to speed up assay-completion time. Sample preparation, which consumes up to two-third fraction of analysis time, plays an important role in almost all biochemical protocols. Dilution and mixing of fluids are two fundamental pre-processing steps in sample preparation. On DMFB architecture, various fluidic operations can be performed on droplets including transport, mixing, splitting, and detection. However, these operations may experience various functional faults (particularly split-errors) during sample preparation, thus impacting adversely the correctness of the application. In particular, volumetric split-errors may occur at any mix-split step of the mixing path due to the inherent randomness of split operations. This poses a significant threat to the reliability of biochemical protocols.

The problem of error-recovery in DMFBs has traditionally been addressed by deploying cyber-physical mechanisms, which are aimed to execute some recovery actions based on the feedback received from on-chip sensors to correct possible volumetric imbalance. Various methods, including checkpointing-based rollback, have been reported in the literature. In this book, the authors have reviewed the existing techniques for achieving error-tolerance including the most recent ones, particularly those suited for reliable sample preparation. New approaches, such as "roll-forward" error correction and sensor-free error-oblivious split operations that lead to the production of target-droplet with correct concentration factors, have been discussed.

Dilution preparation "on-demand" is another important area, where requests for fluid samples with different concentration factors need to be met in real-time. Finally, error management with the most recent class of digital microfluidic biochips known as Micro-Electrode Dot-Array (MEDA) is discussed that guarantees the correctness of the resulting concentration factor without performing any additional rollback or roll-forward action.

Overall, this book is a timely collection of several design optimization techniques that provide the foundation for building error-tolerant microfluidics biochips. Numerous related articles that appeared in journals and conference proceedings in recent years have been diligently reviewed in this monograph. It will serve as a textbook for a graduate course on "Electronic Design Automation (EDA) for emerging technologies", or as a reference book for researchers and industrial practitioners in the area of LoC and EDA. In a nutshell, the book will help bridge the gap between EDA engineers and biologists and inspire them to explore collaboration in this emerging field.

Taiwan,
October 2021

Tsung-Yi Ho, Professor
National Tsing Hua University

Acknowledgments

This book is based on several research articles published by the authors in the area of algorithmic sample preparation with digital microfluidic biochips. We are thankful to IEEE and ACM for granting us copyright permission to use the material that appeared earlier in various journals and conference proceedings. We sincerely thank all the co-authors who contributed to these papers and other related work: Krishnendu Chakrabarty of Duke University, USA; Subhas C. Nandy, Susmita Sur-Kolay, Ansuman Banerjee, Pushpita Roy, and Tapalina Banerjee of Indian Statistical Institute, Kolkata, India; Patha P. Chakrabarti of Indian Institute of Technology Kharagpur, India; Sukanta Bhattacharjee of Indian Institute of Technology Guwahati, India; Sudip Roy of Indian Institute of Technology Roorkee, India; Debasis Mitra of National Institute of Technology Durgapur, India; Sarmishtha Ghoshal and Hafizur Rahaman of Indian Institute of Engineering Science and Technology, Shibpur, India; Subhashis Majumder and Nilina Bera of Heritage Institute of Technology, Kolkata, India; Robert Wille of Johannes Kepler University, Linz, Austria; Tsung-Yi Ho of National Tsing Hua University, Taiwan; Juinn-Dar Huang of National Chiao Tung University, Taiwan. Furthermore, we would like to thank all colleagues of Advanced Computing & Microelectronics Unit (ACMU), Indian Statistical Institute (ISI), Kolkata, and the Department of Computer Science and Engineering, IIT Kharagpur, for their constant support and various services. Financial grants from ISI Kolkata, Indian National Academy of Engineering, CSIR and SERB, Govt. of India, for supporting Microfluidic-CAD research at ACMU are thankfully acknowledged. This work has also partially been supported by Linz Institute of Technology, Govt. of Austria. We would like to thank all fellow lab-mates and research scholars of the NRT Lab, ACMU, who have shared their valuable time and enriched us with their ideas. We extend our sincere gratitude to Prof. Tsung-Yi Ho of National Tsing Hua University, Taiwan, who has kindly agreed to write a foreword for this book. Finally, we are thankful to Boris I. Kharissov of the CRC Press and Gabrielle Vernachio and Allison Shatkin of Taylor & Francis Group for encouraging us to proceed with this book proposal.

Sudip Poddar, Johannes Kepler University, Linz, Austria

Bhargab B. Bhattacharya, Indian Institute of Technology Kharagpur, India

Biographies

Sudip Poddar received the B.Tech. degree in computer science and engineering from the Maulana Abul Kalam Azad University of Technology (formerly known as West Bengal University of Technology), West Bengal, India, in 2008. He received the M.Tech degree in computer science and engineering from the University of Kalyani, India, in 2012. He obtained his PhD degree in Engineering (Computer Science) from Indian Institute of Engineering Science and Technology, Shibpur, Kolkata, India in 2019.

He is currently working as Postdoctoral Fellow in Johannes Kepler University (JKU) Linz, Austria. Prior to joining JKU, he worked as Postdoctoral Fellow in National Taiwan University of Science and Technology (NTUST), Taipei, Taiwan for six months (July 2019-December 2019). He has received Young Career Projects Award (2019) from Linz Institute for Technology (LIT), Govt. of Austria. He is the recipient of Research Associateship (RA) from CSIR (Council of Scientific and Industrial Research), MHRD, Govt. of India (2017-2020). His research interests include computer-aided design for microfluidic lab-on-chip and soft computing. He has authored three book chapters and 16 technical papers that appeared in top-level international peer-reviewed journals and conference proceedings.

Bhargab B. Bhattacharya had been on the faculty of the Computer and Communication Sciences Division at Indian Statistical Institute, Kolkata, since 1982. After his retirement in 2018, he joined the Department of Computer Science & Engineering at Indian Institute of Technology Kharagpur as Distinguished Visiting Professor. He received the B.Sc. degree in Physics from the Presidency College, Kolkata, the B.Tech. and M.Tech. degrees in Radiophysics and Electronics, and the PhD degree in Computer Science, all from the University of Calcutta. His research area includes digital logic testing, and electronic design automation for integrated circuits and microfluidic biochips. He has published more than 400 papers, and he holds ten US Patents. Dr. Bhattacharya is a Fellow of the Indian National Academy of Engineering, a Fellow of the National Academy of Sciences (India), and a Fellow of the IEEE.

Section I

Introduction and Background

1 Introduction

Microfluidics, a technology that enables precise manipulation of small amount of fluid on a tiny chip, has evolved as low-cost and reliable platform for implementing several biochemical protocols, e.g., chemical synthesis, in-vitro diagnostics, drug discovery, and environmental and food toxicity monitoring [138]. Such "micro total analysis systems" (μTAS), also known as "lab-on-chips (LoC)" or "biochips", offer automation, miniaturization, and integration of complex assays, and were fabricated using fluidic components such as microchannels, micropumps, and microvalves, in the nineties. Originally developed by the semiconductor industry, these devices were later used extensively to build a wider class of Micro Electro-Mechanical Systems (MEMS). LoCs can replace expensive and bulky biochemical instruments, and perform clinical diagnosis, massively parallel DNA analysis, protein crystallization, real-time bio-molecular detection, recognition of pathogens, and immunoassays, in much faster and cheaper ways [158]. They are useful for rapid diagnosis of various diseases including malaria, HIV/AIDS, and neglected tropical diseases (NTD) prevalent in developing countries [88, 116, 145, 152]. These devices can also be deployed for providing immediate point-of-care (PoC) health services [139, 143] and for the management of bio-terrorism threats [63, 170]. Microfluidic technology has recently led to the development of synthetic biology and micro-engineered cell culture platforms, known as organ-on-chip (OoC), that mimics *in-vivo* environments of living organs [8]. OoCs offer more realistic ambience for modeling diseases, testing the efficacy of drugs, and for the study of prognosis. The International Technology Roadmap for Semiconductors predicted as early in 2007 that medical science would likely to serve as a major driving force in the future [1]. According to a report released by Research and Markets, the global market revenue of in-vitro diagnostics grew from USD 55.8 billion in 2014 to USD 62.6 billion in 2017 and is expected to reach USD 113.1 billion by 2026 from USD 98.2 billion in 2021, at a CAGR of 2.9% during the forecast period [108], a major share of which can possibly be attributed to the emergence of microfluidic LoCs.

Advancements in the biochip industry have revolutionized biological research and life sciences by bringing a complete paradigm shift. Modern microfluidic devices can handle micro/nano/pico-liter volume of fluids, and a typical one is fabricated as a single chip with only a few square centimeters in area [171]. In general, there are two broad categories of microfluidic devices: (i) static 2D microarray of wells, which is usually used for DNA or protein analysis, and (ii) dynamic chips that emulate a sequence of reactions in a controlled fashion. In the latter class, fluids are enabled to move either through a network of tiny channels (flow-based chips), or as discrete droplets actuated on a surface (digital microfluidic biochips). These chips are likely to replace most of the manual and repetitive laboratory procedures in the near future because of their convenience, versatility, portability, and cost. Most of the basic functions that are needed to execute biochemical procedures such as fluid

DOI: 10.1201/9781003219651-1

transportation, merging, mixing, splitting, reaction, analytic separation, and sensing can be supported on such a biochip with the aid of various actuation mechanisms driven by electrical, pneumatic, thermal, acoustic, or optical means. LoCs can thus automate the execution of complex biochemical protocols, reduce reactant cost, process multiple reactions in parallel, with no or little human intervention, thereby eliminating the burden of doing routine tasks and reducing probable errors. Additionally, biochips efficiently perform various operations of a bioassay at much lower cost and with higher speed [27, 139, 169, 171]. One of the most important applications of the biochip is to automate sample preparation, where the major task is to dilute a fluid sample to the desired concentration factor (dilution assay), or to produce a mixture of multiple fluids in a certain ratio (mixing assay).

In the design cycle of integrated circuits (IC), electronic design automation tools have been widely used during the last three decades. In a similar fashion, with the growing complexities of biochip implementation to encompass the diversity of protocols, and with the widening scope for applications, the need for deploying various computer-aided design (CAD) tools and formal methodologies has been strongly felt while optimizing various objectives such as chip area, the layout-map for fluid-transportation network, module placement, assay-completion time, reactant-cost, and route planning for droplet navigation and contamination wash [52,70,127,164–166,186]. These CAD tools also help chip builders to simulate assays, verify design blueprints, validate expected outcomes in the pre-fabrication stage, and test for possible functional and operational errors that might jeopardize correct behavior of the devices, post-fabrication. The work presented in this book discusses most of the recent techniques used for the mitigation of operational errors that might occur during sample preparation with digital microfluidic biochips.

1.1 BASICS OF MICROFLUIDIC LAB-ON-CHIPS

Dynamic microfluidic biochips, based on the principle of liquid propulsion, are broadly divided into two categories, (i) continuous flow-based biochips (CFMBs), and (ii) digital microfluidic biochips (DMFBs). In CFMBs, fluids are manipulated through permanently etched micro-channels with the help of external pressure sources or integrated mechanical micro-pumps [7, 53]. CFMBs can be used to perform a variety of biochemical assays including polymerase chain reaction (PCR), DNA purification, and protein crystallization. However, they often suffer from various fluidic and channel errors, and therefore, ensuring fault-tolerance during sample preparation poses a significant challenge. DMFBs, on the other hand, use electrical actuation to manipulate discrete fluid packets as carriers to implement various fluidic operations such as dispensing, navigation, merging, mixing, splitting, washing, and sensing at the micro-scale [23, 24, 126]. These chips actuate droplets using software-driven electronic controllers [22]. The salient features of DMFBs lie on the controllability of each individual droplet without the need for using micro-channels, micro-pumps, or micro-valves. Hence, various fluidic operations can be performed on DMFBs in a reconfigurable manner, thereby enabling concurrent execution of multiple applications on a single chip. Thus, DMFBs provide general-purpose programmable microfluidic platforms. Furthermore, error-correcting mechanisms for DMFBs have been well studied in the literature. The detailed architectural descriptions of such biochips along with their applications are described in the literature [23,24,27,39,42,125]. A short review of DMFBs and fluidic operations supported by them are presented in the following section.

1.1.1 DIGITAL MICROFLUIDIC LAB-ON-CHIPS

DMFBs consist of a patterned array of electrodes that can be electrically actuated to perform various fluidic operations [23, 24]. This technology is referred to as "digital microfluidics" since one can manipulate discrete-volume droplets on the electrodes in a "digitized" manner. The top-view of a digital microfluidic biochip, and the execution of basic fluidic operations (dispensing, transporting, mixing, splitting, and sensing) are shown in Fig. 1.1(a) for demonstration. By applying time-varying voltage (low/high) to the electrodes, DMFBs can execute a number of bioassays in parallel [147]. DMFBs contain several microfluidic units including input and waste reservoirs, dispensers, mixers, splitters, sensors, thermal units (heater/cooler), as illustrated in Fig. 1.1(a). For example, we have shown two input reservoirs (used for dispensing the droplets) and one waste reservoir (used to drain unnecessary intermediate droplets from the chip). Sensors, which are used to measure concentration, *pH*, droplet-volume, other physical or chemical properties of fluid droplets, are placed in specified locations on the biochip. Droplet-transportation paths and the location of mixer/splitter modules can be emulated anywhere in vacant regions of a DMFB subject to neighborhood rules, and thus can be dynamically reconfigured on-chip if necessary, e.g., when some electrodes become unusable because of structural degradation or electrical faults.

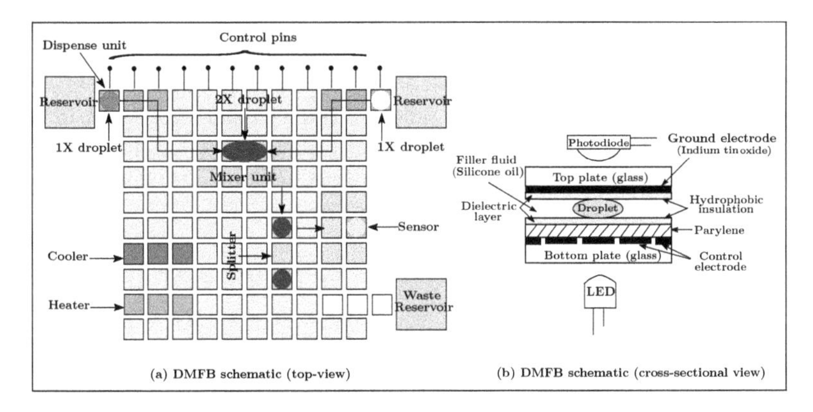

Figure 1.1: DMFB schematic (a) top-view and (b) cross-sectional view

The cross-sectional view of a DMFB cell at the detection site is shown in Fig. 1.1(b). Droplets containing biochemical samples are sandwiched between two parallel glass plates (top and bottom) and placed on a hydrophobic surface over an electrode, as depicted in the figure. The top-plate is covered with a single continuous ground electrode, whereas the bottom-plate is imprinted with an array of controllable electrodes. Silicone oil is used as a filler fluid to fill the gap between the two (top and bottom) plates for preventing droplet evaporation and for reducing surface-contamination. Electrodes are generally connected to control-pins for electrical activation. Light-emitting diodes (LEDs) and photo-diodes are integrated with DMFBs (in fixed positions) for monitoring the status of intermediate or final droplets while executing biochemical assays.

Fluid handling operations such as dispensing, mixing, splitting, and transportation are performed on DMFBs by actuating the droplets in an appropriate fashion using the principle of electrowetting-on-dielectric (EWOD) [126]. Droplets are navigated from one position (source) to another (destination) on the 2-D array when an appropriate sequence of time-varying voltage pulses is applied to control-electrodes to produce a mechanical rolling force along the intended direction. An "electrode actuation sequence" defines the "ON/OFF" switching of an electrode at specific time points. The sequence is determined *a-priori* to execute a given biochemical experiment [61]. Electrode actuation sequences are generally stored in the memory of a microcontroller or a field-programmable gate array (FPGA). A droplet is moved towards an adjacent electrode when an electric field is applied to the junction of two electrodes. This occurs due to the reduction of the interfacial tension between the liquid and the insulator surface (determined according to the Lippman-Young equation) [115]. For example, the control electrode on which a droplet is resting is turned off, and the left electrode is turned on; the droplet will move towards the left (see Fig. 1.1). Note that the droplet must overlap with a neighboring electrode for successful movement towards it. Droplets are usually not allowed to move diagonally in DMFBs. However, all basic fluidic operations can be performed in any location on the biochip by navigating the droplets toward left, right, north, or south, in controlled fashion (Fig. 1.2). Thus, a DMFB acts as a general-purpose programmable microfluidic platform for implementing a large class of biochemical applications. A biochip is connected to a computer via control pins and receives the necessary actuation sequences for executing an assay. A detailed architectural description of DMFBs and their underlying principles have been described in [27,39,125,149]. DMFBs are best suited for a large class of sample preparation, point-of-care clinical diagnosis, proteomics, and immunoassays, among others.

1.1.2 BIOCHIPS BASED ON MICRO-ELECTRODE DOT-ARRAY (MEDA) ARCHITECTURE

A new architecture for digital microfluidic biochips called "Micro-Electrode Dot-Array (MEDA)" has surfaced lately that aims to overcome certain hindrances associated with conventional DMFBs [176,177]. Some of the limitations of DMFBs are: (i) the inability to vary droplet volume in a fine-grained manner because it is fixed and determined by the electrode-area and actuation voltage, (ii) inability to change droplet shape (typically, spherical) if any need arises to facilitate routing through fluidic obstacles, (iii) availability of a few integrated sensors for detecting droplets in real-time, and (iv) stringent requirements concerning fabrication steps. Unlike

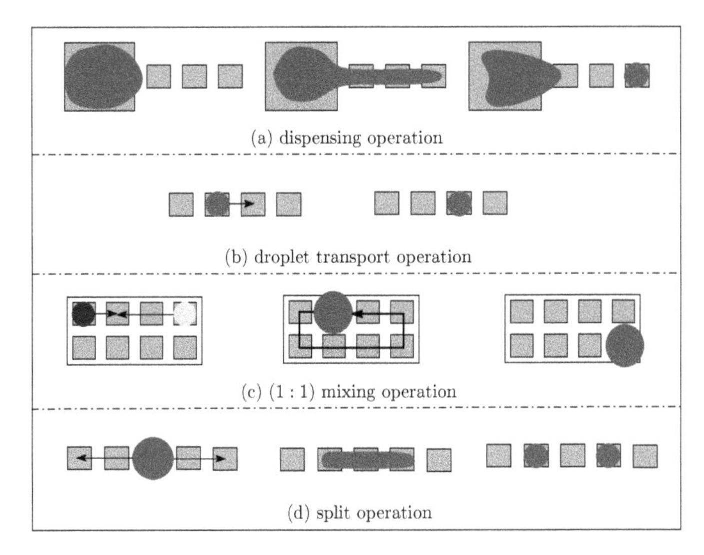

Figure 1.2: Four basic droplet operations on a DMFB device. (a) droplet dispensing, (b) droplet transportation, (c) droplet mixing, and (d) droplet split operation

traditional DMFBs, MEDA chips comprise hundreds of micro-electrodes fabricated using CMOS IC technology, which can be dynamically grouped to form differently-sized functional electrodes and fluidic modules. Thus, droplet volume can be varied suitably with multiple levels of granularity, and their shapes can be morphed so that routing paths can be maneuvered more conveniently. Additionally, each electrode is equipped with an in-built sensor that makes the architecture very conducive to cyber-physical applications, especially those used for error management. Hence, MEDA chips offer several benefits, e.g., real-time sensing, low-cost implementation of assays, flexible droplet routing, ease of system integration with other CMOS modules, and full automation. Note that MEDA inherits routability, scalability, configurability, and portability, all the key factors that enable successful hierarchical design of digital microfluidic biochips. Likewise DMFBs, MEDA chips utilize the principle of EWOD to navigate nanoliter-/picoliter-sized droplets on a two-dimensional array of electrodes, as shown in Fig. 1.3. Such a chip consists of two plates: a top plate and a bottom plate; the top one serves as a single ground electrode similar to that used in DMFBs. However, the bottom plate in MEDA contains a patterned array of hundreds of micro-electrodes (Fig. 1.3(b)). The size of a typical micro-electrode (Fig. 1.3(a)) may be 20 times smaller (e.g., 37μm in length) than that of a conventional DMFB [96, 177].

The architecture of MEDA resembles a two-dimensional "sea of identical micro-electrodes", each of which is called a micro-electrode cell (MC) [177]. Each MC includes a tiny electrode, an activation circuit, and an in-built sensing unit. A high-voltage shielding layer is inserted between a microelectrode and the sensing circuit to ensure the correctness of MC operation. MEDA can efficiently perform all ba-

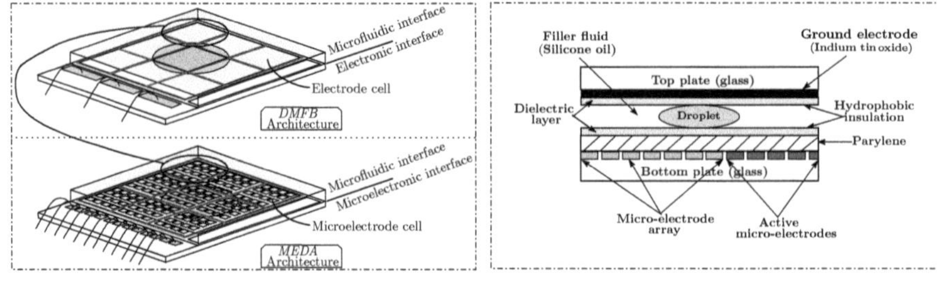

(a) MEDA architecture and micro-electrode cell

(b) MEDA architecture by using scaled-down electrodes

Figure 1.3: (a) Basic architecture of MEDA and micro-electrode cell and (b) MEDA based on scaled-down electrodes (with permission from ACM [119])

sic microfluidic operations such as dispensing, transportation, mixing, and splitting, emulating different micro-components on-chip, but with much deeper and varying levels of granularity. Fluidic modules such as mixer, splitter, diluter, with various functionalities and scale, can be easily composed on MEDA by grouping several micro-electrodes dynamically. Some MEDA-based biochips have been fabricated with TSMC 0.35 μm CMOS technology [177]. The top plate is activated with a 25 V power supply at 1 kHz frequency for performing basic microfluidic operations [85, 92]. Sensing, however, can be accomplished at much higher frequency: 1 MHz using a power-supply of 3.3 V. The volume of a basic micro-droplet is determined by the area of a micro-electrode, and this is the smallest-volume droplet that can be rolled out from the dispenser and processed on MEDA [176]. In MEDA chips, by an "atomic" droplet, we mean the smallest volume of fluid that can be navigated on the chip. Needless to say, the volume of an atomic droplet is determined by the size of a micro-electrode. These atomic droplets can be merged on the chip to form composite droplets of bigger sizes. Thus, it supports different levels of granularity. In order to compare a conventional DMFB with MEDA, let us assume that the area of one electrode of the former is 16X of that of a micro-electrode in MEDA as shown in Fig. 1.4. Let a unit-volume droplet on DMFB be denoted as 1X. Then, on the MEDA architecture, droplets with fractional volumes (*a.k.a.* aliquots) can be formed and different fluidic operations such as transportation, mixing, splitting can be performed on them. This feature greatly enhances the applicability of MEDA chips for reducing reactant-cost during sample preparation, and facilitates droplet routing. For example, the MEDA architecture shown in Fig. 1.4 can compose droplets of volume $\frac{1}{16}$x, $\frac{2}{16}$x, $\frac{3}{16}$x, $\frac{4}{16}$x, $\frac{5}{16}$x, $\cdots$, $\frac{14}{16}$x, $\frac{15}{16}$x, $\frac{16}{16}$x when the size of a 4×4 MEDA micro-electrode sub-array becomes equal to the size of a DMFB electrode. Since MEDA-electrodes are equipped with sensors, they can detect the location (droplet-location sensing) and the characteristics (droplet-property sensing) of droplets in real-time. Such sensing facility enables users to verify the outcome of an on-chip operation in detail and to take necessary corrective actions whenever needed.

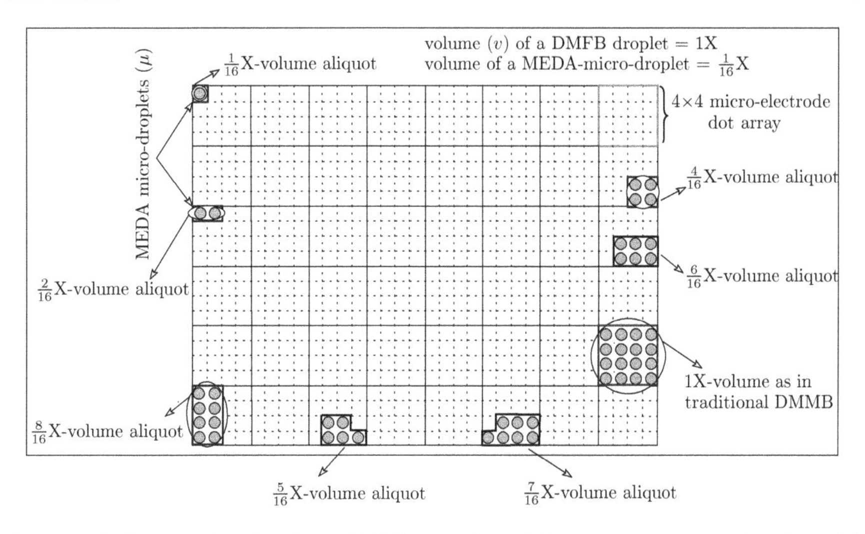

Figure 1.4: Composing droplets of different sizes (aliquots) on a MEDA chip (with permission from ACM [119])

1.2 BASICS OF SAMPLE PREPARATION AND VOLUMETRIC SPLIT-ERRORS

A large number of research papers have been published in the last decade on automated sample preparation using DMFBs [12, 13, 25, 32, 47, 55–57, 60, 92, 99, 100, 113, 163, 185]. Most of them aim to perform a sequence of mixing operations following the (1:1) mixing model for preparing a target mixture. In the (1:1) mixing model, a unit-volume (1X) discrete-droplet with concentration factor[1] (CF) = C_1 is mixed with another unit-volume (1X) discrete-droplet with CF = C_2, to produce a 2X-volume mixture droplet with $CF = \frac{(C_1+C_2)}{2}$. After the mixing operation, it is split into two equal-volume unit-size daughter droplets. Fig. 1.5 demonstrates the (1:1) mixing model implemented in DMFBs. One (1:1) mixing operation and a subsequent split operation are together called a mix-split step (see Fig. 1.5(c)). During sample preparation, a number of reagents $R_1, R_2, \ldots, R_k$, denoted as $\mathscr{R} = \{\langle R_1, c_1\rangle, \langle R_2, c_2\rangle, \cdots, \langle R_k, c_k\rangle\}$, are mixed with a ratio of $\{c_1 : c_2 : \cdots : c_i : \cdots : c_k\}$, for preparing a target mixture, where c_i denotes the CF of R_i, $0 \leq c_i \leq 1$ for $i = 1, 2, \ldots, k$, and $\sum_{i=1}^{k} c_i = 1$ (to ensure the validity condition of mixing ratio) [163]. Because of the inherent attributes of (1:1) mixing model supported by a DMFB platform, each c_i, depending on user-specified error-tolerance limit τ ($0 \leq \tau < 1$), is approximated as $\mathscr{R} = \{\langle R_1, c_1 = \frac{x_1}{2^n}\rangle, \langle R_2, c_2 = \frac{x_2}{2^n}\rangle, \cdots, \langle R_k, c_k = \frac{x_k}{2^n}\rangle\}$; where $\sum_{i=1}^{k} x_i = 2^n$ and $x_1, x_2, \ldots, x_k, n \in \mathbb{N}$ [163]. Note that for a given τ, we need to choose the minimum value of $n \in \mathbb{N}$ such that each c_i in mixture $\mathscr{R}$ is approximated as $\frac{x_i}{2^n}$, where $x_i \in \mathbb{N}$, subject to $\max_i\{|c_i - \frac{x_i}{2^n}|\} < \tau$ and $\sum_{i=1}^{k} x_i = 2^n$ [163]. Most of the sam-

[1] If sample X of volume x is diluted with a buffer solution B of volume y, then the CF(X) is defined as: x/(x + y); 0 < CF < 1; CF of raw sample (buffer) is assumed to be 1(0)

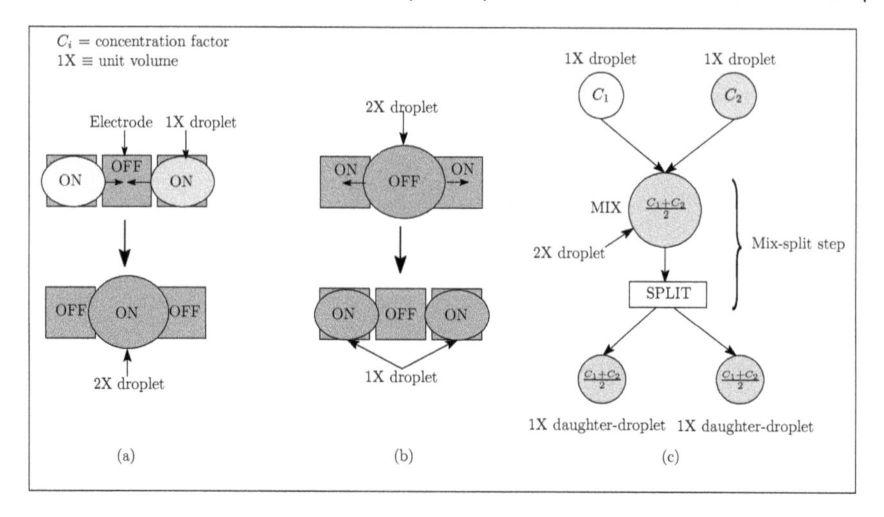

Figure 1.5: Schematic of (1:1) (a) mixing, (b) splitting of droplets, and (c) a mix-split step on a DMFB platform

ple preparation algorithms [131,163] produce a sequence of (1:1) mix-split steps, and the sequence is envisaged as a *mix-split tree/graph*, which depicts the mix-split steps that are needed to reach the target ratio starting from input reagents (see Figs. 2.2 and 2.3).

It has been observed that several run-time fluidic errors may occur while executing an assay on a DMFB which badly affect expected outcomes. These operational or functional errors may occur even if the biochip passes the tests for manufacturing defects or structural faults (e.g., imperfections in electrodes, electrical opens/shorts, electrode degradation, etc.) Most of these run-time errors are caused at the time of dispensing or splitting operations, causing a change in the ideal volume of a droplet. During sample preparation, a sequence of mix-split operations needs to be performed to achieve the desired ratio of their constituent fluids. Incidentally, the concentration factor of an individual component in a fluid mixture is very sensitive to the volumes of participating droplets, which are used in the assay. Thus, any volumetric error caused during dispensing or splitting operation adversely affects the concentration profile of the final mixture. Also, there are inherent uncertainties regarding the occurrence of these errors during the execution of the assay. In sample preparation, several fluids (known as source) are supplied as input, and at the end of the dilution or mixing assay, the desired solution (known as target) is obtained. Source-droplets are generally injected on the biochip from input reservoirs, target droplets are routed to destination reservoirs, and waste droplets are drained into waste reservoirs. All these reservoirs are usually situated around the boundary of the chip. Ideally, a 1X-volume droplet should be dispensed from the source-reservoirs. However, due to some defects, in practical scenario, a larger/smaller droplet of volume $(1+\varepsilon)X/(1-\varepsilon)X$, $0 < \varepsilon < 1$, may be dispensed from the reservoirs as shown in Fig. 1.6(b)–(c). This type

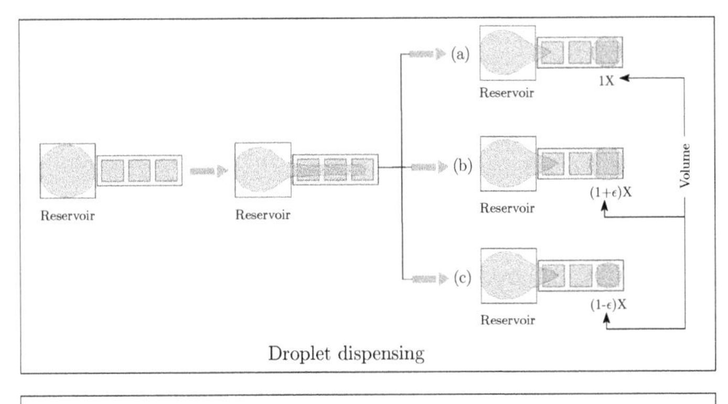

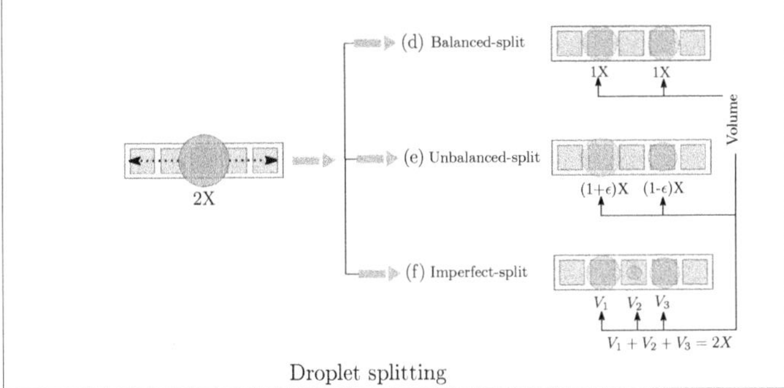

Figure 1.6: Illustration of droplet-dispensing and droplet-splitting operations in DMFBs

of error can be corrected by simply returning the erroneous droplet back to the source reservoir and re-dispensing again. Such schemes do not waste any droplets, and can correct the errors at the cost of re-dispensing time. Moreover, other accurate droplet-emission mechanisms are also known for dispensing precise volume of droplets from the reservoirs [27, 46]. For simplicity, we assume that a dispense operation from a reservoir is error-free. Recovery becomes more complicated when volumetric split-errors occur because their impact on target concentrations is not properly understood. In ideal scenario, a 2X-volume droplet should be split into two 1X-volume daughter-droplets. However, due to physical-defects or fluidic-uncertainties, a mother-droplet is sometimes split into two unequal-volume daughter-droplets of size $(1+\varepsilon)$ and $(1-\varepsilon)$ as shown in Fig. 1.6 (e). It may also sometime be divided into three parts as shown in Fig. 1.6 (f). Experimentally it has been observed that unbalanced split-errors occur most of the time ($\approx 80\%$) following split-operations and such an error may cause up to 7% volume-variation in the daughter-droplets [122, 129].

In recent years, several cyber-physical methods have been proposed to address the problem of error-recovery in DMFBs [3, 4, 58, 105, 192]. Traditional methods perform error-recovery operations as follows: During the execution of a bioassay, intermediate droplets are sent to designated spots on the chip where sensors are located. These droplets are discarded as waste if any error is detected by the sensors. The relevant portion of the assay is then re-executed by rolling back to the previous checkpoint where error-free droplets had been saved. Although rollback schemes can successfully recover the assay from split-errors, they suffer from the following inherent limitations:

- Those methods require either an online re-synthesis scheme for regenerating the electrode actuation sequence dynamically or additional memory in which pre-computed re-synthesis solutions for error-recovery have to be stored.
- A rollback procedure increases the completion time of the bioassay significantly. Accordingly, they are not suitable for applications such as flash chemistry, where reactions need to take place in a very short time. For example, the reaction time of SwernMoffatt-type oxidation is about 0.01 seconds at 20°C [187].
- Prior error-recovery methods assume that "re-execution is always possible". However, enough copy-droplets may not always be available to enable re-execution.
- Rollback based on pre-computed re-synthesis solutions requires some kind of dictionary, which, in turn, requires additional memory – contradicting the goal of designing low-cost biochip devices for field deployment and point-of-care testing in resource-constrained environments.
- Earlier approaches also fail to provide any guarantee on the number of rollback attempts, i.e., how many rollback iterations will be required in order to correct the error. Hence, the error-recovery process becomes non-deterministic in time.
- Cyber-physical DMFB platforms may contain a limited number of sensors (due to cost constraints). As a result, they may introduce additional latency for error-detection, since each resultant droplet needs to be routed to a sensor location for error-detection.
- Prior cyber-physical approaches require $O(MN)$ time for searching appropriate copy-droplets (when an error occurs) in an $(M \times N)$ size biochip while recovering from an error [58, 105, 106, 192]. Thus, they increase the assay-completion time in the presence of large number of errors.

In other words, most of the previous work relies on on-chip sensors, and a controller is needed for monitoring the status of intermediate mix-split operations. Also, the corresponding sensor signals should be fed back to the controller in a timely manner for recovery actions.

1.3 SCOPE OF THE BOOK

As mentioned beforehand, the presence of volumetric split-errors on DMFBs adversely affects the accuracy of sample preparation. In the following chapters of the book, we look into the error recovery problem in detail, present a survey of existing rollback-based methods, and discuss newer approaches for the management of split-errors, which rely on "roll-forward" techniques and the property of "error obliviousness" perceived during mix-split steps. We also present a technique for error-tolerant sample preparation suitable for implementation with MEDA biochips.

1.4 ORGANIZATION OF THE BOOK

- Chapter 2: We introduce the basics of mixing models that are supported by microfluidic biochips (DMFB and MEDA biochips). Next, we present a survey of existing methods that are deployed for sample preparation with DMFBs. The effects of volumetric split-errors on the concentration profile of target droplets are discussed.
- Chapter 3: In this chapter, we discuss the basics of various error-recovery methods suitable for DMFBs and MEDA biochips that are based on roll-back techniques.
- Chapter 4: This chapter describes the problems of prior error-recovery methods with cyber-physical DMFBs and presents a deterministic solution to error-correcting sample preparation.
- Chapter 5: A rigorous analysis of the effects of multiple volumetric split-errors on the concentration profiles of target-droplets, is presented here.
- Chapter 6: We introduce the concept of error obliviousness during mix-split operations and discuss how this idea translates into the production of target solutions accurately without the need for any sensor or feedback mechanism.
- Chapter 7: We discuss the problem of multi-target sample preparation "on-demand", i.e., when the desired concentration factors of target solutions are not known in advance. Various optimization issues and techniques for split-error management in this context have been presented.
- Chapter 8: We describe here how the power of MEDA biochips can be harnessed to implement error-tolerant sample preparation without using any droplet split-operation, thereby eliminating the source of all such errors.
- Chapter 9: Finally, we summarize the content of this book and discuss possible future research directions in the area of error-tolerant sample preparation.

2 Background

Diluting a fluid with a buffer and mixing of multiple fluids in a given ratio are two major tasks of sample preparation (SP), and these steps are indispensable in almost all biochemical assays/protocols [6, 74, 128, 163, 185]. A significant amount of time is spent to collect, transport, and prepare samples with various concentration factors (*CF*) for further processing [44]. In SP, two or more biochemical fluids are to be mixed in a certain volumetric ratio, and a sequence of mix-split operations is generally performed on the constituent fluids to achieve the desired target-ratio. Traditional test-tube-based laboratory methods for SP may require a long time for measuring fluid volumes and enabling the mixing process with high accuracy of *CF*s. Moreover, a large number of medical applications mandate very fast processing of pathological samples. For example, the mortality rate of a patient suffering from sepsis shock out of bacterial infection in the bloodstream increases by 7.6% every hour [82]. Automation of SP with DMFBs mitigate most of these drawbacks by performing fluidic steps such as merging, splitting, mixing, and dispensing, very quickly on a chip. Thus, LoCs greatly aid for rapid completion of biochemical experiments at low cost and prevent the delay in diagnostics and treatment.

The set of input reagents used in sample preparation is denoted as $\mathscr{R} = \{\langle R_1, c_1\rangle, \langle R_2, c_2\rangle, \langle R_3, c_3\rangle, \ldots, \langle R_k, c_k\rangle\}$[1], where $\sum_{i=1}^{k} c_i = 1$ and $0 \leq c_i \leq 1$ for $i = 1, 2, 3, \ldots, k$. A ratio is approximated as $\{R_1 : R_2 : R_3, \ldots, R_k = x_1 : x_2 : x_3 : \ldots : x_k\}$, where $\sum_{i=1}^{k} x_i = N^n, N, n \in \mathbb{N}$ [15]. The underlying microfluidic mixing model determines the value of N (e.g., N is 2 for the (1:1) mixing model), whereas the value of n is determined by a user-specified error-tolerance limit τ ($0 \leq \tau < 1$). The accuracy of the target-*CF* becomes higher for lower values of τ. After approximating a mixing ratio, SP-algorithms create a directed acyclic graph called *mix-split graph/tree*, based on the optimization goals such as the minimization of input reagents, waste production, or the number of mix-split operations [131, 163].

2.1 PRELIMINARIES

2.1.1 MIXING MODELS

In DMFBs, mix-split operations during SP are generally performed using unit-size droplets. In $(l : m)$ mixing model, l unit-volume droplets are mixed with m unit-volume droplets producing a mixture droplet of volume $(l + m)$ after mixing operation. Based on the volume of the input droplets during mixing operations, the following mixing models are used in DMFBs: (i) $l = m = 1$, (ii) $l \neq m$, and (iii) $l = m > 1$. For example, various mixing models supported by regular DMFB and MEDA biochips are shown in Fig. 2.1. DMFBs usually support the mixing of two equal unit-volume droplets of different concentrations through a (1:1) mixing model

[1] c_i denotes the concentration factor (*CF*) of R_i

DOI: 10.1201/9781003219651-2

(see Fig. 2.1 (a)). This is due to the architectural constraints of DMFBs which facilitate fluidic operation with fixed (equal) volume droplets. On the other hand, MEDA biochips allow fractional aliquoting of droplets (i.e., they can handle droplets with different volumes) in a fine-grained manner. Hence, MEDA supports various mixing models as shown in Fig. 2.1 (a)–(b).

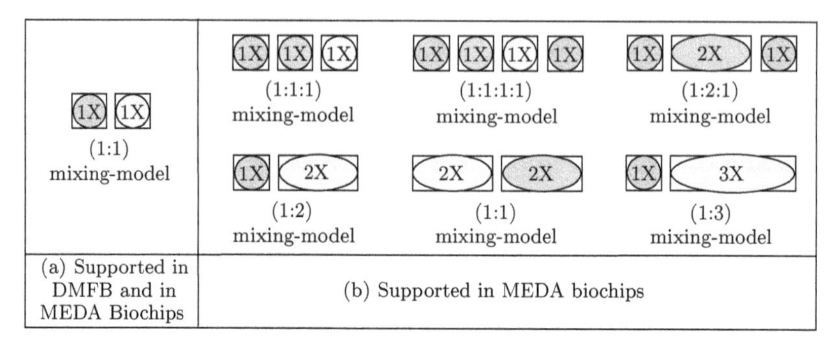

Figure 2.1: Mixing-models supported in DMFB and MEDA biochips

Till date, most of the work on sample preparation in microfluidic biochips are based on the (1:1) mixing model since it is the easiest to implement [12, 13, 25, 32, 47, 55–57, 60, 92, 99, 100, 113, 163, 185]. In the (1:1) mixing model, two fluids, one being a mixture of input fluids with a ratio $\{R_1 : R_2 : R_3 : \cdots : R_k = x_1 : x_2 : x_3 : \cdots : x_k\}$ and another with a ratio $\{R_1 : R_2 : R_3 : \cdots : R_k = y_1 : y_2 : y_3 : \cdots : y_k\}$ are mixed with same volumetric ratio, and after the mixing operation, split into two-equal volume droplets with concentration ratio $\{R_1 : R_2 : R_3 : \cdots : R_k = \frac{x_1+y_1}{2} : \frac{x_2+y_2}{2} : \frac{x_3+y_3}{2} : \cdots : \frac{x_k+y_k}{2}\}$. In dilution, two input fluids (commonly known as sample and buffer) are mixed, where the ratio $\{\text{sample} : \text{buffer} = x : y\}$ can be represented with the concentration factor (CF) of the sample, i.e., $\frac{x}{x+y}$ where $x + y = 2^n$, n is a non-negative integer, and is chosen depending on the required accuracy of CF. The denominator in CF is chosen as 2^n for convenience in the domain of digital microfluidics, particularly for the (1:1) mixing model.

2.1.2 CONCENTRATION AND DILUTION FACTORS

In sample preparation, the term concentration factor (CF) is defined as the amount of raw sample fluid present in a target droplet ($0 \leq CF \leq 1$) [163]. In other words, CF is the ratio of initial volume of the sample to the final volume of the target mixture (i.e., the reciprocal of the dilution factor (DF), $CF = \frac{1}{DF}$). Therefore, both CF and DF are defined as the volume-to-volume ratios, and the value of CF (DF) is always becomes less (greater) than or equal to 1.

2.1.3 AUTOMATED DILUTION OF A SAMPLE FLUID

Dilution is an essential preprocessing step required in almost all biochemical protocols for producing a desired ratio of two fluids [102], one of which is usually a buffer fluid. Dilution is needed in real-time PCR for cDNA, immunoassays for detecting cytokines in serum samples or in Trinder's reaction, where a *DF* of 200 or more is used [129]. Dilution is commonly used in biological studies for creating a number of stock solutions with different *CFs* [51, 74, 131]. Additionally, dilution gradient (diluting a sample over a certain range of *CF*s) plays an important role in drug design, toxicity analysis, clinical diagnostics, to name a few. It is also needed in the quantification of test results.

As mentioned earlier, dilution of a biochemical sample/reagent is a special case of mixing two input fluids (sample and buffer) with a certain volumetric ratio corresponding to the desired target concentration factor. A sample fluid with $CF = C_1$ can be diluted by mixing it with another sample of the same fluid with $CF = C_2$, where $C_2 < C_1$. When these two fluids are mixed with $(l : m)$ volumetric ratio, *CF* and volume of the resulting droplet become: $\frac{l \times C_1 + m \times C_2}{l+m}$ and $l+m$, respectively. A subsequent balanced-split operation produces two daughter droplets each with volume $\frac{l+m}{2}$. In dilution, a raw sample is mixed with a special type of fluid (neutral to the sample) termed as diluent or buffer, e.g., water or other liquid. Therefore, the concentration of the buffer solution is always viewed as 0% ($CF = 0$), and the *CF* of the raw sample fluid is viewed as 100% ($CF = 1$).

2.1.3.1 Linear and serial dilution

Often a sample fluid with a wide range of *CF*s is required in many real-life biochemical experiments [17, 28, 113, 160], e.g., multiple *CF*s of a sample are required in bacterial susceptibility test satisfying certain "gradient pattern". In drug design, it is important to determine the minimum amount of an antibiotic that inhibits the visible growth of bacteria isolate (defined as Minimum Inhibitory Concentration (MIC)). The drug with the least concentration factor (i.e., with the highest dilution) that is capable of arresting the growth of bacteria is considered as MIC. Therefore, in sample preparation, generating dilution gradients over linear and non-linear ranges is of interest. These types of *CF*s can be produced using linear dilution and serial dilution. Recently, a number of methods have been proposed for microfluidic LoCs to generate certain gradient patterns [13, 29, 73, 73, 87, 161, 174, 180].

Linear dilution that dilutes a stock solution (sample) into a linear range of *CF*s over $CF = 0$ (0%) to $CF = 1$ (100%), offers sensitive tests [183]. In linear dilution, *CF*s of a sample fluid generally appear in arithmetic progression, e.g., 5%, 10%, 15%, 20%, 25%. A linear dilution of *CF*s of sucrose ranging from 10% to 40% is required in the case of sucrose gradient analysis [17]. Serial dilutions dilute a stock solution into a logarithmic (or, equivalently, exponential) scale. A serial dilution is the stepwise dilution of a sample fluid. In this type of dilution, *CF*s are generated by repeating the dilution step, using the previous *CF* as input to the next dilution. *CF*s are generated in geometric progression since dilution-fold (mixing-ratio) is same in each step. For example, the following *CFs* are generated in serial dilution: $\frac{1}{5}$, $\frac{1}{25}$, $\frac{1}{125}$, $\frac{1}{625}$. Serial dilutions are widely used to accurately create highly diluted solutions for biochemical experiments.

2.1.3.2 Exponential and interpolated dilution

Serial dilution can be divided into two categories depending on the nature of the two input fluids: exponential (or logarithmic) dilution and interpolated dilution. In exponential dilution, a sample fluid is repeatedly mixed (diluted) with a buffer fluid with the same volumetric ratio. As a result, the *CF* of a sample fluid decreases exponentially by a factor of 2 after each mixing step (cycle). Hence, *CF* of the sample fluid becomes $\frac{C}{2^n}$ when a (1:1) mix-split operation is recursively performed n times (dilution factor is 2). For example, when a raw sample with $CF = \frac{128}{128}$ is mixed with a buffer with $CF = \frac{0}{128}$ using (1:1) volumetric ratio, the *CF* of the mixture droplet becomes: $\frac{64}{128} = \frac{1}{2}$. However, when a sample fluid with $CF = C_1$ is mixed with another dilution of the same sample with $CF = C_2$, in (1:1) volumetric ratio, then *CF* of the resultant droplet becomes $\frac{C_1+C_2}{2}$, where $0 < C_1, C_2 < 1$ (assuming that the same buffer is used as diluent in both cases). This type of dilution is called interpolated dilution. For example, when a sample with $CF = \frac{120}{128}$ is mixed with another sample of the same fluid with $CF = \frac{60}{128}$ using (1:1) volumetric ratio, the *CF* of the resultant mixture becomes: $\frac{90}{128} = \frac{45}{64}$.

2.2 PRIOR-WORK ON SAMPLE PREPARATION

Thies *et al.* [163] first proposed a mixing algorithm called *MinMix*, for mixing two or more fluids in a given volumetric ratio with a microfluidic biochip. For the (1:1) mixing model, *Minmix* first represents the target-*CF* of each input fluid x_i as an n-

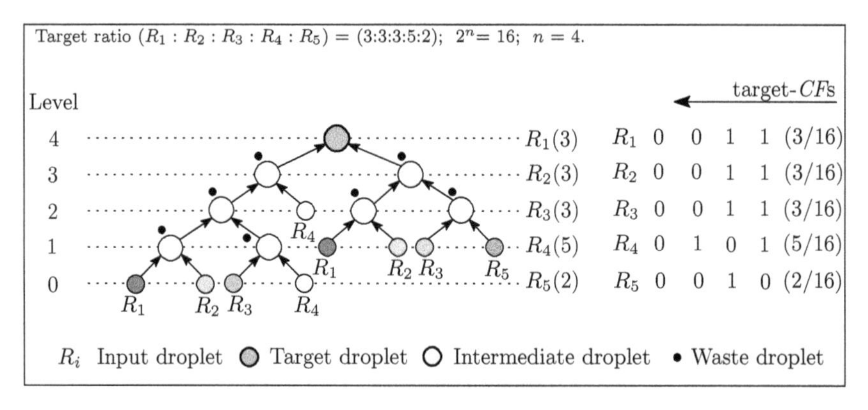

Figure 2.2: Mixing tree of depth 4, created by *MinMix* [163], for the target ratio (3:3:3:5:2)

bit binary-string, where, n is a user-defined integer that is related to the accuracy of *CF*s. Then, it scans the bits of the (0,1) matrix of size $k \times n$, column-wise, from right-to-left, to create a binary mixing-tree of height n. The mixing tree is constructed after level-wise merging of two leaf-nodes (or, one leaf-node and one non-leaf node, or two non-leaf nodes) where the merged node indicates the mixture of the corresponding two input fluids or intermediate fluids. When a digital microfluidic platform is used, each leaf-node in a mixing tree corresponds to a 1X-volume input droplet, and a non-leaf (internal) node represents a 1X-volume mixture droplet obtained after performing a (1:1) mix-split operation on them. Note that the sequence of (1:1) mix-split operations that are required for producing the desired target-mixture of k fluids (denoted by the root of the mixing tree) can be obtained by a post-order traversal of the mixing tree. This method generates the desired target-droplet with the minimum number of (1:1) mix-split operations when no sharing of waste droplets that are produced after every mix-split step. An example binary mixing-tree for generating the target-ratio $\{3:3:3:5:2\}$ of five different fluids is shown in Fig. 2.2. The rightmost bits in the 4-bit binary representations of R_1, R_2, R_3, R_4, and R_5 are 1, 1, 1, 1, 0, respectively; hence a reagent droplet each for R_1, R_2, R_3, R_4 is required at level 0.

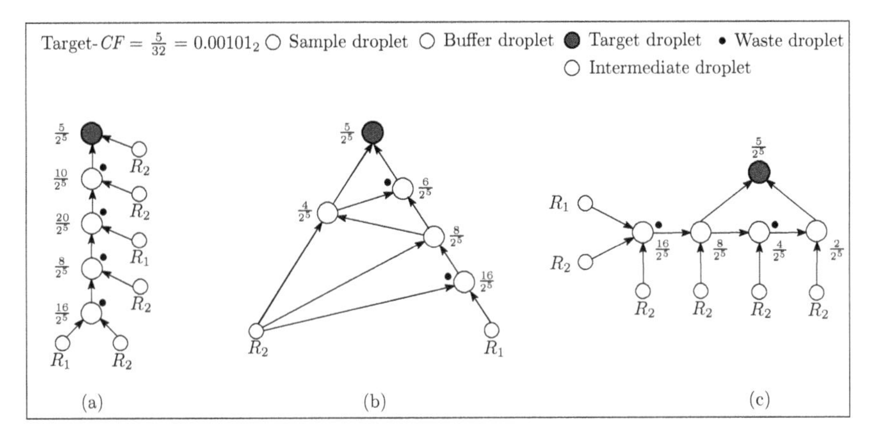

Figure 2.3: Dilution of a sample using (a) *twoWayMix* [163], (b) *DMRW* [131], and (c) *REMIA* [60]

Algorithm *twoWayMix* [163], which is a special case of *MinMix*, creates a mix-split tree/graph for generating a particular dilution of a sample. In this case, the number of reagents is two: sample and buffer. As before, the desired target-*CF* is approximated as an n-bit binary fractional number, ignoring all the 0's on the right of the least significant 1 bit. In the first step, it mixes one sample droplet and one buffer droplet in (1:1) ratio. Next, it scans the remaining bits from right-to-left to decide whether the sample droplet or buffer droplet is to be used in the next (1:1) mix-split operation. This method generates a dilution tree of depth n in $O(n)$ time and bounds the error in *CF* of the target-droplet by $\frac{1}{2^{n+1}}$. An example mixing tree generated by *twoWayMix* for the target-$CF = \frac{5}{2^5}$ is shown in Fig. 2.3(a). Note that the total number of 1's (0's plus one) in the binary fraction of the target-*CF* denotes the number of required sample (buffer) droplets needed for producing two target-droplets. Algorithm *twoWayMix* does not require any on-chip reservoir for storing the intermediate droplets since it always mixes the current intermediate droplet with the sample/buffer droplet in the next mix-split step. However, it produces $n' - 1$ waste droplets when $n' (n' \leq n)$ mix-split operations are required to generate the target-*CF*.

In order to reduce waste-droplets generated in the intermediate mix-split steps, an algorithm based on binary-search technique, called '*Dilution and Mixing with*

Reduced Wastage' (*DMRW*) was proposed by Roy *et al.* [131]. It mixes the droplets generated in the last two intermediate mix-split operations. This method starts by setting sample and buffer as upper and lower bounds, respectively. Next, in each iteration, it updates the lower (upper) bound with the mean of the two if the target is larger (smaller) than the mean. The process terminates when the target-droplet is reached. *DMRW* requires some extra storage space (for storing intermediate droplets) since it shares intermediate droplets for reducing the number of waste droplets. In order to reduce the dilution time, a specialized rotary mixer based on $(m : m)$ mixing model was also proposed $(m \geq 1)$. Note that for the above-mentioned example, compared to *twoWayMix*, *DMRW* reduces waste-droplets from 4 to 2 (see Fig. 2.3(b)). Roy *et al.* later proposed another algorithm namely, '*Improved Dilution/Mixing Algorithm*' (*IDMA*) for further improving the performance of *DMRW*.

Reducing the overall reactant usage is also an important parameter in sample preparation. In this direction, Huang *et al.* first proposed '*REactant MInimization Algorithm*' (*REMIA*) for diluting a sample to a desired target concentration factor [60]. Unlike prior work, fluids obtained by serial dilution of the raw sample are also used here to obtain the target droplet. Thus, it performs the dilution process in two steps: i) interpolated dilution and ii) exponential dilution step. In the interpolated step, it builds a mixing tree whose leaf-nodes are a subset of exponential *CF*-values, such that reactant usage is minimized. It also creates an exponential dilution tree via serial dilution for producing fluid *CF*s required at the leaf-nodes of the mixing tree. Fig. 2.3(c) shows the overall dilution process for target-$CF = \frac{5}{32}$ using *REMIA*. Another method called '*Graph-based Optimal Reactant Minimization Algorithm*' (*GORMA*) was proposed for minimizing reactant usage during sample preparation [25]. Roy *et al.* presented'*Generalized Dilution Algorithm*' (*GDA*) for preparing a target-*CF* using a set of input stock solutions with arbitrary *CF*s [134].

Multiple concentration factors of the same sample are often required in many bioassays. For example, some reagents are required in multiple *CF*s in bacterial susceptibility tests, satisfying certain "gradient" patterns, e.g., linear, exponential, or parabolic. Although single-target sample-preparation algorithms can be used to generate multiple target-*CF*s, they increase reagent-cost due to repeated execution of several dilution steps. It becomes more complex when multiple droplets of different *CF*s need to be prepared at the same time. In this direction, Bhattacharjee *et al.* proposed '*Pruning Based Dilution Algorithm*' (*PBDA*) for generating multiple dilutions of a given sample [12]. Based on the single-target sample-preparation algorithm (*REMIA*), Huang *at al.* introduced '*WAste Recycling Algorithm*' (*WARA*) for reducing reactant-usage during multi-target sample preparation [100]. On the other hand, for optimizing the number of mix-split steps, Mitra *et al.* proposed '*Multiple Target Concentration*' (*MTC*) algorithm for preparing multi-*CF*s of a sample without using any on-chip storage for intermediate droplets [113]. Many other algorithms were proposed in recent years for preparing multi-target-*CF*s [14, 32, 120, 121]. Table 2.1 summarizes the key features of various sample-dilution algorithms for DMFB-platforms based on their target-profile and optimization objectives (reduction of mix-split steps/reactant-droplets/waste-droplets).

Table 2.1
Scope of various sample-preparation algorithms.

Dilution algorithm	Suitable for LoC type	For Single/ multiple targets?	Optimization objectives: Reduction of		
			#Mix-splits	Waste	Reactant-cost
twoWayMix [163]	DMFB	single-target	yes	no	no
DMRW [131]	DMFB	single-target	no	yes	no
IDMA [135]	DMFB	single-target	yes	yes	no
REMIA [60]	DMFB	single-target	yes	no	yes
GORMA [25]	DMFB	single-target	yes	no	yes
PBDA [12]	DMFB	multi-target	yes	yes	no
RSM [57]	DMFB	multi-target	yes	yes	no
WARA [100]	DMFB	multi-target	yes	yes	yes
MTC [113]	DMFB	multi-target	yes	yes	yes
Dinh *et al.* [31]	DMFB	multi-target	yes	yes	yes
Poddar *et al.* [121]	DMFB	multi-target	yes	yes	yes
Bhattacharjee *et al.* [14]	DMFB	multi-target	no	yes	no
MRCM [98]	MEDA	multi-target	yes	yes	yes

2.3 EFFECT OF VOLUMETRIC SPLIT-ERRORS ON TARGET CONCENTRATION

Prior approaches to sample preparation with DMFBs assume that every balanced mix-split step correctly produces two 1X-volume daughter droplets after splitting a 2X-volume mother droplet. However, the outcome of a split operation depends on various parameters such as channel gap, droplet size, degree of contact angle induced by the phenomenon of electrowetting-on-dielectric (EWOD), and switching time of actuation voltages. As discussed earlier in Section 1.2, erroneous fluidic operations may occur in a biochip even if it is deemed to be defect-free after manufacturing tests [23, 159]. In addition, the volume of a fluid droplet may also change unexpectedly due to several parametric faults such as non-uniform electrode coating, the application of unequal actuation voltages, or asynchronous switching of actuation signals on the two electrodes that touch the mother droplet from two opposite sides. Such imperfections may arise due to dispense error or unbalanced-split operation that may occur during droplet dispensing or splitting, thus affecting the volume of droplets [27,40,75]. In general, a split may be balanced, unbalanced, or imperfect as shown in Figs. 1.6 and 2.4. Balanced split produces two equal-volume 1X droplets after a split operation, whereas during unbalanced split, two unequal-volume droplets are produced. An imperfect split may leave a small residue on the middle electrode, apart from producing two equal/unequal volume droplets on the neighboring electrodes. It has been experimentally observed that the accuracy of sample preparation

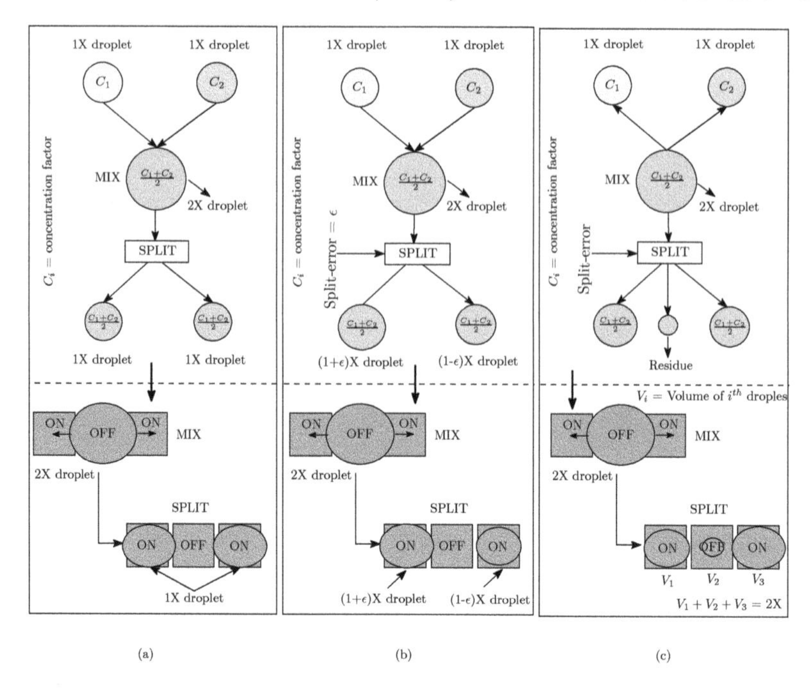

Figure 2.4: Illustration of (1:1) mix-split operation: (a) balanced split, (b) unbalanced split, and (c) imperfect split, on a DMFB platform

suffers due to "after-split" volumetric imbalances in most of the cases (approx 80% of all errors), and these errors may cause up to 7% volumetric split-error [129]. Note that each mix-split operation may be a potential source of such errors and thus, may affect the outcome after the execution of an assay. An example of preparing target-$CF = \frac{87}{128}$ for accuracy level 7 is shown in Fig. 2.5 when 7% volumetric split-error occurs in mix-split Step 6. We observe that even a single volumetric split-error affects the target-*CF* badly causing an error ($=\frac{1.39}{128}$) in the target-droplet, which exceeds the allowable error-tolerance limit = $\frac{0.5}{128}$ (see Fig. 2.5). Moreover, the *CF*-error in the target-droplet may also increase rapidly when multiple mix-split steps suffer from split-errors. Therefore, the reliability of the target-*CF* strongly depends on the correctness of droplet-splitting mechanisms. Since ideal split operations may not be realizable in practice, one needs to develop various error-correcting strategies to ensure accurate sample preparation.

2.4 ERROR-CORRECTION DURING MULTI-TARGET SAMPLE PREPARATION

Reliability (i.e., accuracy), in the context of multiple-target sample preparation, is of utmost importance in several real-life biochemical applications [185]. For example, in many experiments such as DNA sequencing and protein crystallization,

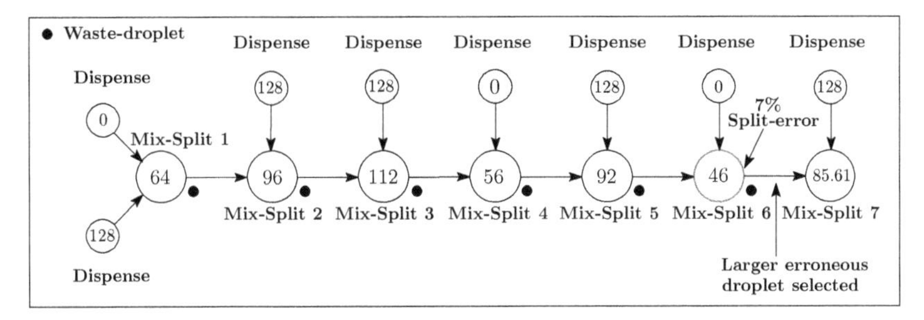

Figure 2.5: Effect of volumetric split-error on target-$CF = \frac{87}{128}$, when the larger-size daughter-droplet is used in the next step

hundreds of solutions need to be prepared accurately [58]. Sample preparation also plays a major role in diagnostics. For example, a blood sample needs to be diluted with Phosphate Buffered Saline (PBS) for the measurement of complete blood cell (CBS) count for point-of-care applications [50]. A variety of concentrations (linear and non-linear dilutions) are required in cytotoxicity assays (routinely performed in high-throughput systems), for which mixing needs to be performed precisely [174]. In this context, the reliability of sample preparation becomes crucial in situations that involve mission-critical or life-critical biochemical assays such as pathological diagnostics, DNA analysis, or drug design, where the correctness of assay outcomes cannot be compromised.

We observe that most of the existing algorithms on multiple-target sample preparation attempt to share the intermediate droplets (produced after intermediate mix-split steps) for reducing reactant-cost and assay time. As a result, the underlying mix-split tree, which represents the sample-preparation process based on the (1:1) mixing model, turn out to be a graph [57, 100]. Therefore, improvising error-correction mechanisms for multi-target sample preparation becomes far more challenging because of the complexities present in the mix-split graph.

2.5 CONCLUSION

This chapter highlights the need for error-tolerant sample preparation that is discussed at full length later in this monograph. A summary of various algorithms and related tools for sample preparation with digital microfluidic biochips is presented. These CAD-tools can be efficiently used to optimize different objective functions such as reactant-cost, waste production, or the number of mix-split steps while preparing desired dilution or mixtures of fluids. We have demonstrated with an example how the accuracy of target concentration factors can be degraded due to the presence of volumetric split-errors.

Section II

Literature Review

3 Error Recovery Methods for Biochips

As discussed earlier, an LoC offers a very convenient platform for automating clinical diagnostics [143], environmental monitoring [191], and drug discovery [39], among many other applications. Concomitantly, with the increasing complexities of LoCs and to cope with diverse optimization goals, the need for Computer-Aided Design (CAD) tools has been strongly felt while synthesizing various classes of digital microfluidic biochips. Although significant research has already been carried out over the last decade on the automated design of DMFBs [21, 24, 153, 154, 157], most of them are restricted to off-line synthesis, and they do not address any robustness or reliability issues in real-time. Several fluidic errors may often crop up during droplet manipulation due to various defects, imperfect operations, aging, contamination, or chip degradation. Such malfunctioning may render the diagnostic outcome of microfluidic-based medical kits unreliable [106]. In numerous experiments, highly accurate and precise fluid handling operations are required, e.g., in drug discovery and DNA analysis. Modern LoCs thus aim to integrate fluid-handling operations, assay execution, sensing, error-detection, and error-recovery mechanisms on a single platform. In consequence, error-recoverability has evolved as one of the most crucial steps in the design of DMFB-based analytics and diagnostic devices.

3.1 DESIGN OBJECTIVES FOR ERROR-RECOVERY

In recent years, a number of research articles have been published in the area of error-recovery in DMFBs [3, 4, 58, 105, 106, 192]. These recovery methods vary in their capabilities with respect to error coverage, overhead, utilization of available resources, and time-to-response. Needless to say, fast, efficient, and cost-effective error-recovery is a challenging task as there exists trade-off among a number of parameters such as response time, area overhead, control memory usage, on-chip storage requirement, sensor availability, and scalability (multiple-error handling) [65].

For implementing error-recovery methods on DMFBs, one or more on-chip sensors are needed for detecting functional errors in real-time. Once an error is flagged, the control software invokes appropriate recovery actions such as re-executing some portions of the assay with copy-droplets that are stored at earlier checkpoints, or the re-synthesis of protocols requiring new placement of fluidic modules and/or rerouting of droplet-pathways. The time needed to take care of such correcting actions is called *error-recovery-time*, which is an inverse measure of system responsiveness. Error-recovery-time is the sum of (i) triggering-time (T_t): time duration between the occurrence and detection of an error, and (ii) execution-time (E_t): time required to perform error-recovery operation (where an error occurs). Also, triggering-time depends on: (i) traveling time (Tr_t): time required to move a droplet from its current

DOI: 10.1201/9781003219651-3

location to a sensor location (Fig. 3.1), and (ii) error-detection time (Ed_t): time taken by the on-chip sensor to detect the error. Moreover, error-detection time depends on the technology used in the sensor. For example, triggering-time becomes negligible when Charge-Coupled Device (CCD)-based camera is used for error-detection (Tr_t becomes negligible) [105, 106]. However, the light from a CCD camera may influence chemical substances, for example, fluorescent markers in the droplet [105]. Therefore, other electronic or electro-chemical sensing systems are preferred when photosensitive samples or reagents are used in the experiment, and these devices may be little slower. For sensing transparent droplets with an optical detector, some coloring agents may need to be mixed with them [192].

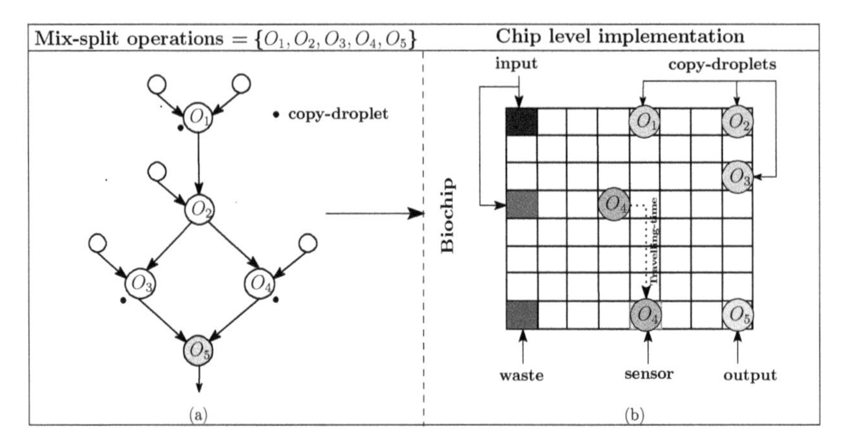

Figure 3.1: Illustration of an assay being executed on a digital microfluidic biochip

The waste-droplets produced during intermediate operations, if sensed to be error-free, can be stored on-chip (copy-droplets) for future use (see Fig. 3.1). These droplets can be utilized for performing error-recovery operations. Error-recovery-time reduces when a large number of copy-droplets are stored on the chip. However, they may create additional routing constraints to other active droplets, thus delaying the completion time of the bioassay. Note that some operations may not supply any copy-droplet, e.g., when all daughter-droplets produced by a split operation are utilized in subsequent steps. For example, mix-split operation O_2 is unable to provide any copy-droplet as shown in Fig. 3.1. There also exists some trade-off between control memory usage and error-recovery-time. Memory utilization depends on the underlying error-recovery scheme. Control-memory may be used to store anticipated error-recovery operations, *a-priori*. However, this approach requires sufficient memory for storing a large number of error-recovery operations. On the other hand, *posteriori* approaches (which do not require any prior information for error-recovery) may reduce the cost of memory usage.

Regardless of what recovery methods are used, handling multiple fluidic errors is still a challenging task. For example, *a priori* approaches [106, 192] can manage recovery from a single fluidic error but may fail to produce an efficient solution when multiple fluidic errors occur at the same time. *Posteriori* approaches [5, 105] can tackle well multiple errors but at the cost of higher assay-completion time.

Lukas [103] developed a static simulator for visualizing the set of fluidic operations currently executing on a chip. The simulator shows the effect of an error on droplet-volume. Grissom *et al.* [49] developed a software framework for visualizing the sequence of droplet mix-split steps on DMFBs. However, the problem of error-recovery in the context of sample preparation was not studied earlier. Prior work on cyber-physical DMFBs only focused on the development of error-recovery methods for general assays. They are based on either a rollback and re-execution strategy, or the usage of pre-stored error-recovery sequences [3, 4, 58, 105, 106, 192]. In recent years, several error-recovery methods have also been developed for Micro-Electrode-Dot-Array (MEDA)-based biochips [92, 94, 193]. In the following sections, we will discuss, in detail, error-recovery methods for sample-preparation algorithms implemented on DMFBs and MEDA biochips.

3.2 ERROR RECOVERY WITH REGULAR DMFBs

3.2.1 INTEGRATED CONTROL-PATH DESIGN AND ERROR RECOVERY

Zhao *et al.* addressed the problem of error-recovery that occurs during on-chip execution of biochemical assays with DMFBs [192]. They described a synthesis method including control-paths, and an embedded error-recovery mechanism. In this method, the best locations for inserting fluidic checkpoints[1] are estimated during bioassay execution, and an error-recovery subroutine is designed for each of them. The subroutine includes all fluidic operations (e.g., dispensing, transportation, mixing, and sensing) from the immediate predecessor checkpoint along all paths in the sequencing graph to the current checkpoint. During execution, the method checks the correctness of the intermediate droplets at each checkpoint. When an error is detected, the erroneous-droplets are diverted to the waste reservoir, and the corresponding error-recovery routine is invoked in the subsequent cycle. During the execution of an assay, a checkpoint can only be reached when no error is detected at all its predecessor checkpoints. This means that an error is always localized between the current checkpoint and the immediate predecessor checkpoints. Any error can thus be corrected simply by re-executing the appropriate error-recovery subroutine. However, the method performs corrective actions in a standalone manner, i.e., while executing the error-recovery subroutine, it stops all other ongoing operations of the assay.

[1] A checkpoint in an assay is defined as a prior time-stamp in the mix-split graph for which the intermediate (copy) droplets are stored on-chip (if found free from fluidic errors) so that they can later be used for error recovery if needed.

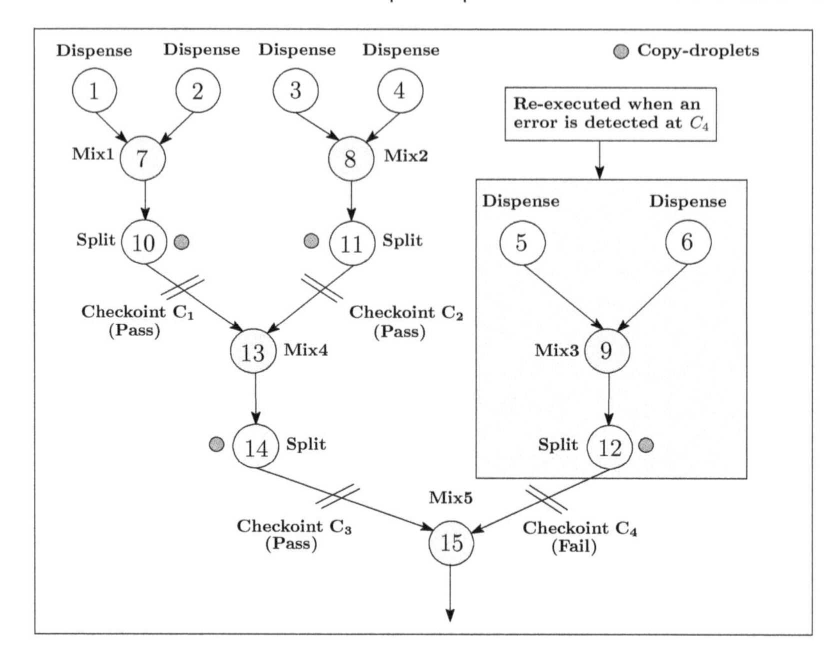

Figure 3.2: Checkpoint insertion and re-execution subroutine

Fig. 3.2 shows an example of assigning checkpoints at various stages of an assay. It inserts four checkpoints C_1, C_2, C_3, and C_4 for monitoring the outcome of intermediate fluidic operations. A re-execution sub-routine is assigned to checkpoint C_4 for error-recovery. It re-executes the fluidic operations included in this routine when an error is detected at C_4.

3.2.2 SYNTHESIS OF PROTOCOLS ON DMFBs WITH OPERATIONAL VARIABILITY

Alistar *et al.* [3] proposed a fault-tolerant synthesis scheme for digital microfluidic biochips, where the initial sequencing graph is converted into a fault-tolerant sequencing graph (FSG) that captures possible faulty scenarios. It addresses the effects of volumetric split-errors under certain fault-models, and uses capacitive sensors to detect whether the daughter-droplets suffer from any volumetric-errors following every split-operation. When a sensor detects an error (error in droplet-volume exceeds a given threshold limit), erroneous droplets are sent back to the source location (where the split was performed) for re-merge and re-split operation. For k split-errors, it requires maximum $k+1$ re-merge and re-split operations to correct a split-error as shown in Fig. 3.3. It performs sensing operation k times except the last split operation, since the last split-error does not change the value of target-CF during sample preparation.

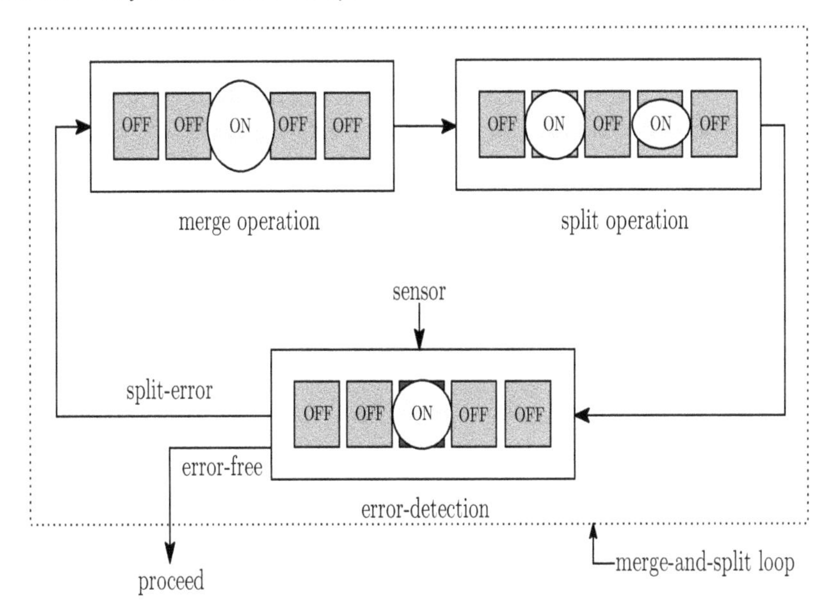

Figure 3.3: Merge-and-split loop for each mix-split operation of the assay

3.2.3 ERROR RECOVERY IN CYBER-PHYSICAL DMFBs

Lue *et al.* [105] proposed a method for error-recovery by placing sensors at various checkpoints to dynamically reconfigure the biochip. It uses a cyber-physical-based re-synthesis strategy to compute electrode actuation sequences for error-recovery operations, with minimum impact on time-to-response. In order to handle an error, each fluid-handling operation is classified into two categories: (i) reversible operation: this includes splitting and dispensing and (ii) non-reversible operation: this includes dilution and mixing. A reversible split-error can be corrected by re-merging the two erroneous daughter-droplets and splitting them again. The dispensing errors are corrected by returning the erroneous droplets back to the source reservoir, and re-dispensing them.

For non-reversible errors, the method requires some additional cost for recovery. It attempts to use copy-droplets for correcting non-reversible errors. If the required copy-droplet is found, it is moved to an appropriate position of the biochip to initiate the error-recovery process. However, it may need to re-execute a number of predecessor operations depending on the availability of the copy-droplets. In the worst-case scenario, it may need to perform error-recovery operations at the cost of high assay-completion time.

3.2.4 DICTIONARY-BASED REAL-TIME ERROR RECOVERY

In order to reduce response time and enable applications to flash chemistry, Luo *et al.* [106] proposed a hardware-assisted error-recovery method with cyber-physical DMFBs. The method computes error-recovery actuation sequences for possible errors that may occur while executing the assay, and stores the corresponding re-synthesis solutions (actuation sequences to perform recovery actions) into the memory of a microcontroller or a field-programmable gate array (FPGA), akin to a dictionary. It adapts two data compaction techniques from the literature[2] for reducing its size. When an error occurs, it simply looks up the kth entry of the error-dictionary for performing error-recovery (without doing online re-synthesis) using appropriate copy-droplets. The method reduces error-recovery time by storing necessary actions in the controller memory in the form of a decision tree.

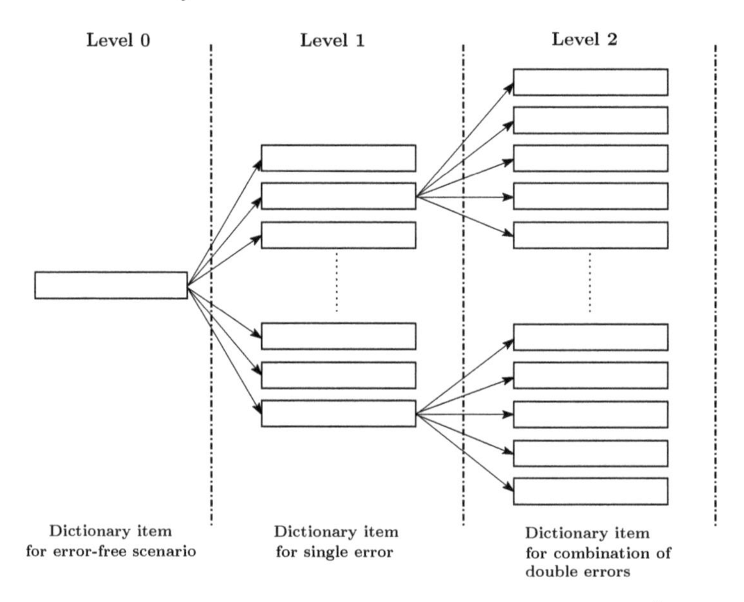

Figure 3.4: Tree-like structure of the error-dictionary. Entries at the kth level provide the re-synthesis solution for possible combinations of k errors [106]

Note that the error-dictionary may have multiple levels, and an entry in a particular level represents the re-synthesis solution for possible combinations of errors (see Fig. 3.4). For x operations and k errors, the number of entries in the dictionary at k^{th} level becomes $\binom{x}{k}$. The size of the dictionary becomes excessively large when x and k increase. In order to meet the trade-off between the response time and memory usage, it considers only up to two erroneous operations in a bioassay.

[2] COO (coordinate list) compaction [9] and run-length encoding [173].

3.2.5 DYNAMIC ERROR RECOVERY DURING SAMPLE PREPARATION

Hsieh *et al.* [58] proposed an optimization algorithm and discussed associated chip design for sample preparation with DMFBs, which includes architectural synthesis, layout synthesis, and dynamic error-recovery. In this method, the checkpoints are assigned to the mixing graph based on some predefined parameter p $(0 \leq p \leq 1)$ defined by the architectural design. It next performs scheduling and resource binding for fluidic operations and on-chip resources. The former determines the execution order (start-time and end-time) for each fluidic operation so as to reduce sample-preparation time. Resource binding maps each fluidic operation (mixing, sensing, or storing of copy-droplets) of the mixing graph to a particular on-chip resource (mixer, sensor, or storage units) when needed.

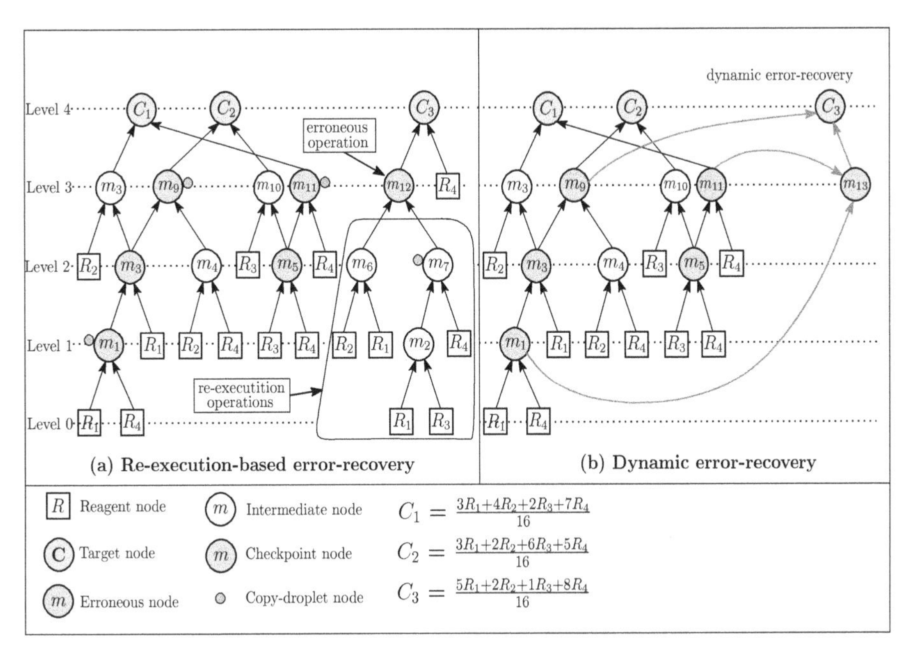

Figure 3.5: Example of (a) re-execution-based error-recovery and (b) dynamic error-recovery [58]

This method attempts to recover errors using stored copy-droplets (if available) instead of re-executing all associated operations. In order to minimize the error-recovery time, it re-generates intermediate droplets closest to the affected target-*CF* of the mixing graph. In the worst case, it re-executes all required operations. Examples of re-execution-based error-recovery and dynamic error-recovery are shown in Fig. 3.5.

3.2.6 REDUNDANCY-BASED ERROR RECOVERY IN DMFBs

Alistar *et al.* proposed an online technique for error-recovery [4, 5] based on space or time redundancy. In time redundancy, re-execution of erroneous operations is invoked, and in space redundancy, redundant-droplets are created by executing the same operations in parallel before the occurrence of the error. The method chooses the right combination of redundancy so that fault-tolerance is maximized, satisfying the assay-completion deadline.

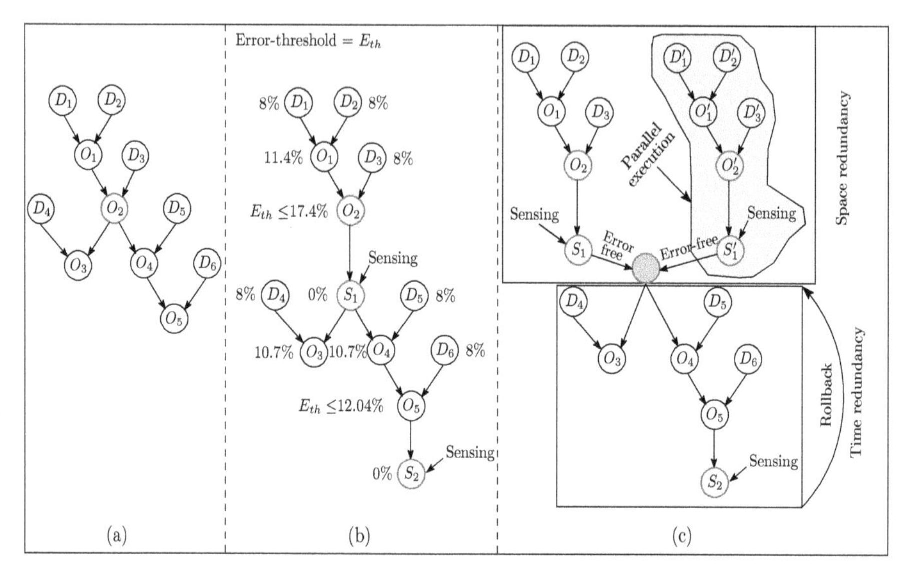

Figure 3.6: Example of (a) an initial sequencing graph, (b) a sequencing graph with sensing operation (after error-estimation), and (c) online redundancy assignment in the sequencing graph for error-tolerance [5]

Checkpoints are inserted at different nodes of the mix-split graph based on error-propagation analysis [162]. It assigns the checkpoints dynamically by adjusting the error-threshold limit at runtime, depending on the current fault status. Space redundancy is adopted when enough chip area is available, otherwise it deploys time redundancy for error-recovery. Error-recovery time is reduced by utilizing stored copy-droplets while performing recovery operations. Note that there exists trade-off between space and time redundancy. The former needs more on-chip storage and functional resources, which may compromise droplet routability (increased transportation delay) when an error is detected. The latter performs rollback operations, and in turn, increases assay-completion time.

As an example, we show an initial sequencing graph in Fig. 3.6(a). The error-limits for all fluidic operations are also shown in Fig. 3.6 (b). Note that the error-limit of an operation is calculated from its "intrinsic error limit"[3], and the limits of input

[3]Error-range of each fluidic operation that captures droplet-volume variation in the worst-case.

operations. In order to check whether or not an error has actually occurred at runtime, the method invokes sensing operation after O_2 and O_5, as the calculated error-limits (17.4% and 12.04%) exceed the pre-defined error-threshold limit = 12% (see Fig. 3.6(b)). Initial assignments of redundancy operations (space and time) to the sequencing graph are shown in Fig. 3.6(c). It has been observed that the method adopts space redundancy to tolerate transient errors up to operation O_2, i.e., it performs the same operation in two parallel paths at runtime and takes the correct droplet between the two tested droplets (one droplet from each path). It adopts time redundancy operation when both droplets become erroneous after O_2, the error being detected by sensor S_2.

3.3 ERROR-RECOVERY WITH MEDA BIOCHIPS

3.3.1 DROPLET SIZE-AWARE AND ERROR-CORRECTING SAMPLE PREPARATION

A hybrid error-recovery approach during sample preparation with MEDA biochips has been developed by Li *et al.* [92]. In order to tackle the inherent uncertainty of fluidic operations (e.g., splitting operation), this method utilizes *MEDA*-specific fluidic operations, such as fine-grained control of droplet sizes and real-time droplet sensing operations. The approach proposed in this work combines both rollback [192] and roll-forward [122] techniques for error-recovery. Note that almost all prior-work on error-recovery used a rollback approach for error-correction, where a set of mix-split operations is re-executed for error-recovery using back-up (copy) droplets (stored on on-chip reservoirs). These types of error-recovery approaches discard the erroneous droplets whenever an error is detected. On the other hand, in the roll-forward approach [122], the erroneous droplets are combined together either to cancel the effect of split-errors or to reduce the concentration error when the target-droplet is reached.

In the method developed by Li *et al.* [92], errors in concentration factors (CFs) are classified into two categories, namely, major and minor. In-built sensing capability of MEDA electrodes is used to determine the difference (error) between the measured (actual) CF and the reference (expected) CF-value of each droplet in real-time. A CF-error is said to be minor if it does not exceed the allowable error-tolerance range; otherwise, it is called a major-error. Minor errors are ignored, and the following recovery actions are invoked only when a major error is detected. Two droplets D_i and D_j are said to be paired-droplets if D_i is mixed with D_j in the mixing graph, i.e., D_i (D_j) is a pair-droplet of D_j (D_i). For example, droplets D_1 and D_2 in Fig. 3.7 (a) are called paired-droplets because they are mixed to produce two-droplets of D_3. The volume of paired-droplets are sensed in real-time on MEDA while executing the dilution graph. If a major-error is detected in one of the paired-droplets, then the situation is termed as single error scenario (SES). Otherwise, it is called paired-error scenario (PES).

The management of SES is explained with an example (Fig. 3.7 (a)–(d)). Suppose a major-error is detected in droplet D_1, and the size of the detected droplet is 0.8X instead of being 1X. Next, the cost of the error-recovery is estimated as follows:

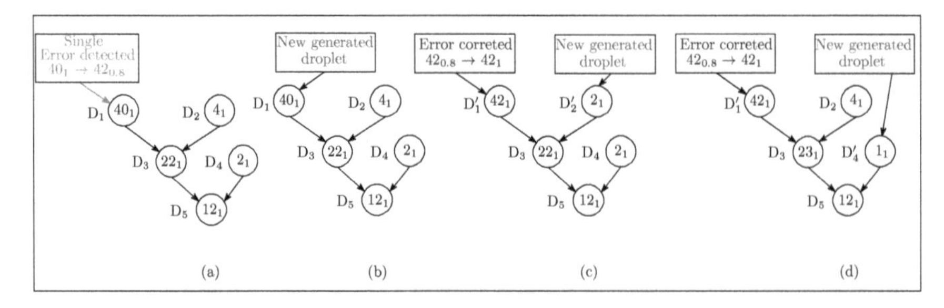

Figure 3.7: Example depicting (a) the original dilution graph and updated dilution graph while performing, (b) first recovery approach, (c) second recovery approach, and (d) third recovery approach for single error

(i) regeneration of D_1 to replace the erroneous-droplet D'_1, (ii) utilization of D'_1, i.e., correcting the volume of the erroneous droplet D'_1 by performing a droplet-regulation operation. Accordingly, a new droplet D_2 is generated for producing the droplet D_3 with the desired *CF*, and (iii) utilization of both droplets D'_1 and D_2 in the error-recovery process as shown Fig. 3.7 (d). In this case, a new droplet D'_4 is generated for obtaining droplet D_5 with the desired *CF*. Finally, a suitable error-recovery technique is chosen based on the following criteria: to achieve (i) as first objective: reduction of sample cost, (ii) as second objective: reduction in the number of mix-split operations, and (iii) as third objective: reduction in the total number of droplets that are being generated. In the case of PES, major errors are detected in both droplets D_1 and D_2 as shown in Fig. 3.8 (a). These are corrected following a similar procedure. The criteria include the cost of regenerating the droplet pair to replace the erroneous ones, or the overhead needed to correct volumetric errors in the pair, or the cost of all extra droplets that are required to be produced.

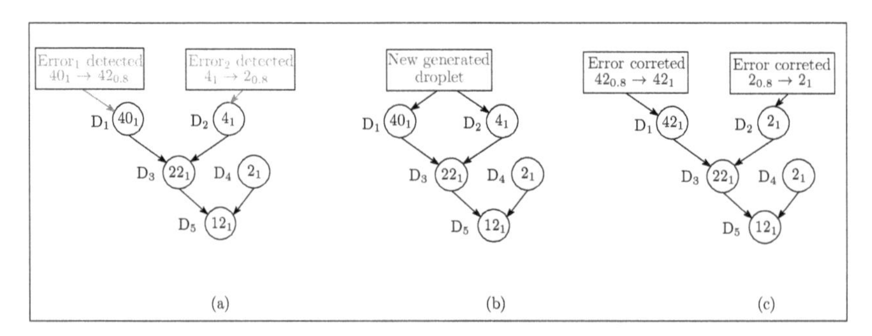

Figure 3.8: Demonstration of the (a) original dilution graph and updated dilution graph while performing, (b) first recovery approach, and (c) second recovery approach for PES

3.3.2 ADAPTIVE ERROR RECOVERY IN MEDA BIOCHIPS

In another work [94], Li *et al.* introduced Probabilistic-Timed-Automata (PTA)-based error-recovery. The status of droplets is monitored in real-time using the in-built sensors on micro-electrodes, and this information is used to dynamically reconfigure the biochip in the presence of some errors. Splitting errors are classified into three categories (i) major error, (ii) minor error, and (iii) no error, and appropriate corrective actions are taken.

In order to classify the split-errors, the volumes of two daughter-droplets are measured after each split operation, and their volumetric difference is computed. The following expression is then used to classify the split-error into the above three categories.

$$EF_{sp} = 1 - \frac{|NOE_{d1} - NOE_{d2}|}{max\{NOE_{d1}, NOE_{d2}\}} \tag{3.1}$$

where NOE_{d1} is the number of microelectrodes occupied by one daughter-droplet and NOE_{d2} denotes the same for the other daughter-droplet. Finally, based on two user-defined threshold values Th_1 and Th_2, split-errors are categorized as: (i) major if $EF_{sp} \leq Th_1$; (ii) minor if $Th_1 < EF_{sp} \leq Th_2$; (iii) no error if $EF_{sp} > Th_2$.

In this method, the correctness (uniformity) of mixing operations is estimated by detecting the permittivity of each droplet. A software tool is used for mapping different ranges of permittivity levels into distinct colors. Based on the observed colors of the mixture, imperfections in mixing operations are categorized into major, minor, no-error in real-time. In the case of uniform/homogeneous mixing (i.e., no error), the droplet following a mixing operation will contain only one specific color (defined as final color). Otherwise, it may show up multiple colors associated with the visualization of the mixed droplet. The following expression is used to quantify the error-factor EF_{mo} for a mixing operation:

$$EF_{mo} = \frac{NOE_{fc}}{NOE_d} \tag{3.2}$$

where NOE_{fc}, NOE_d represent the number of microelectrodes highlighted with the final color in the measurement window and the total number of microelectrodes occupied by the entire droplet, respectively. Similar to split error, it uses two threshold values Th_3 and Th_4 to classify mixing errors into: (i) major if $EF_{mo} \leq Th_3$; (ii) minor if $Th_3 < EF_{mo} \leq Th_4$; (iii) no error if $EF_{mo} > Th_4$. Experiments suggest that these threshold values should be set as follows: Th_1 = 0.50, Th_2 = 0.80, Th_3 = 0.70, and Th_4 = 0.90. The dilution operation is said to commit a major error when the outcome of either mixing or splitting induces a major error. When both operations suffer from a minor error, the outcome of dilution leads to a minor error.

In this method, a local error-recovery procedure is handled to recover from mixing, splitting, or dilution errors. It adopts different flows to recover from various types of errors instead of invoking the same rollback-based error-recovery procedure. For example, a mixing operation with minor error is recovered by repeating it on the same mixer a certain number of times. In the case of major error, the

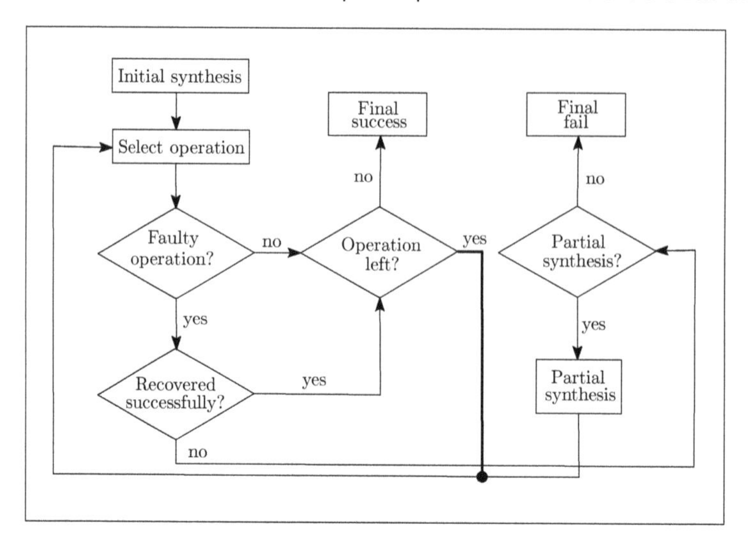

Figure 3.9: Overall control flow for error-recovery

same mixing operation is performed in another mixture. While performing error-recovery operations, stored back-up droplets are utilized for reducing the overhead of error-recovery. Splitting errors (major and minor) are corrected simply by re-doing merge and split operations a fixed number of times. However, recovery operations are deemed to have failed if an error reappears after making several attempts. In the case of dilution operations, the aforementioned recovery actions are taken depending on whether mixing or splitting is affected by an error. Fig. 3.9 depicts the overall flow of the error-recovery process.

3.3.3 ROLL-FORWARD ERROR RECOVERY IN MEDA BIOCHIPS

An error-recovery method has been proposed by Zhong *et al.* [193] that exploits the inherent granularity, sensing scheme, and sophisticated fluidic operations supported on MEDA biochips. In particular, the power of "droplet aliquoting", i.e., the availability of fractional-volume droplets is utilized in order to recover from split errors. Recovery from mixing errors is accomplished based on a "predictive analysis model". A roll-forward technique is also used to reduce error-recovery time.

Droplet aliquoting is an advanced fluidic operation offered by MEDA-biochips, which allows deriving a smaller droplet from a larger one. This operation is not feasible in regular DMFBs due to architectural constraints. The above-mentioned method utilizes droplet aliquoting in an adaptive way to recover from erroneous splitting as shown in Fig. 3.10. For example, suppose a split operation produces two unequal volume daughter-droplets D_1 and D_2, the size (S) of D_1 being larger than that of D_2. In order to equalize their volumes, a small aliquot droplet D_c can be extracted from D_1 and merged with D_2. Thus, the volumetric difference between them can be reduced to an acceptable range. The example shown in Fig. 3.10 demonstrates this procedure.

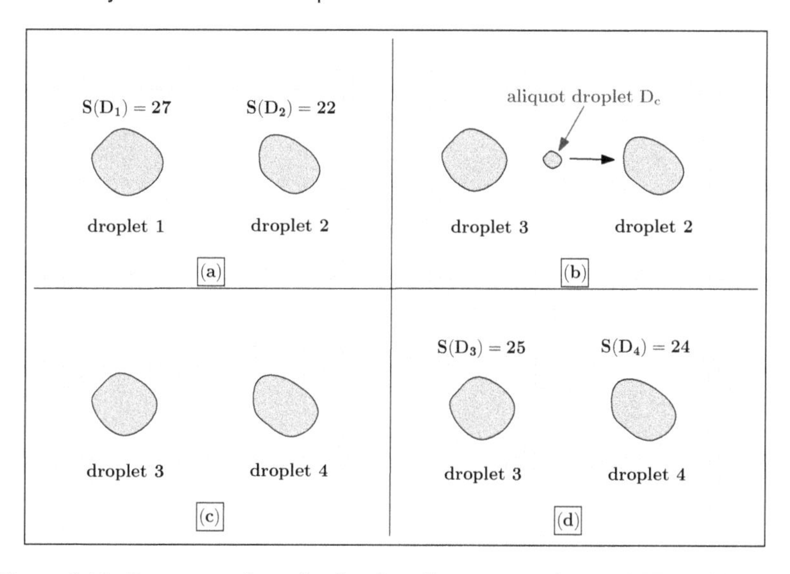

Figure 3.10: Demonstration of a droplet-aliquot operation on MEDA biochips

Initially, the volumetric difference between D_1 and D_2 was 5. Next, a small aliquot droplet D_c (two microelectrodes in size) is extracted from D_1 and merged with D_2 to reduce the imbalance in their volumes (Fig. 3.10 (b)). As a result, the size of the new droplets D_3 and D_4 become 25 and 24, respectively (Fig. 3.10 (c)). This method takes full advantage of droplet aliquot operations, and the overall control flow for error-recovery from splitting errors is shown in Fig. 3.11.

In addition, the proposed method uses Expression 3.3 to recover from errors introduced by mixing operations [86]. In other words, it captures the relationship between mixing rate versus elapsed time on MEDA biochips (for a droplet with 7μL volume). The dependence of the mixing coefficient on time is given by:

$$C_{mix}(t) = 1 - e^{-\frac{t}{\lambda}} \tag{3.3}$$

where $C_{mix}(t)$ denotes the extent to which mixing has been completed at time t and λ is a time constant. A higher (lower) value of λ denotes slower (faster) mixing. The overall control flow for recovering from a mixing error is shown in Fig. 3.12. The controller first checks whether or not the mixing operation is erroneous. In the case of an error, the mixing rate is calculated using Expression 3.3. Next, it predicts whether the droplet can achieve the desired completion rate. If so, the mixing operation is executed on the same mixer again. Otherwise, when the mixing rate is lower than its average value, the droplet is moved to a different location for mixing (i.e., for error-correction). Note that a mixing operation is attempted at most five times (to put a bound on error-recovery time) as many biochemical samples degrade very quickly, e.g., fibronectin degrades within 10s [45].

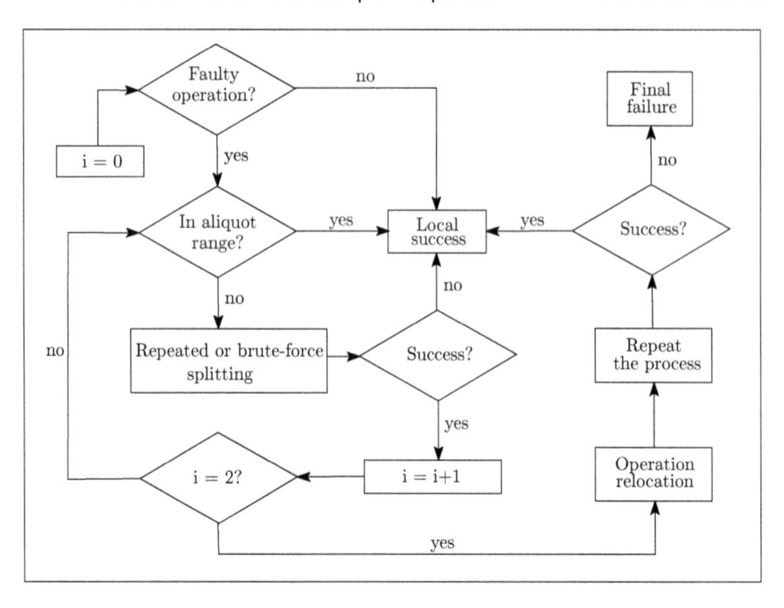

Figure 3.11: Error-recovery control flow for splitting operations

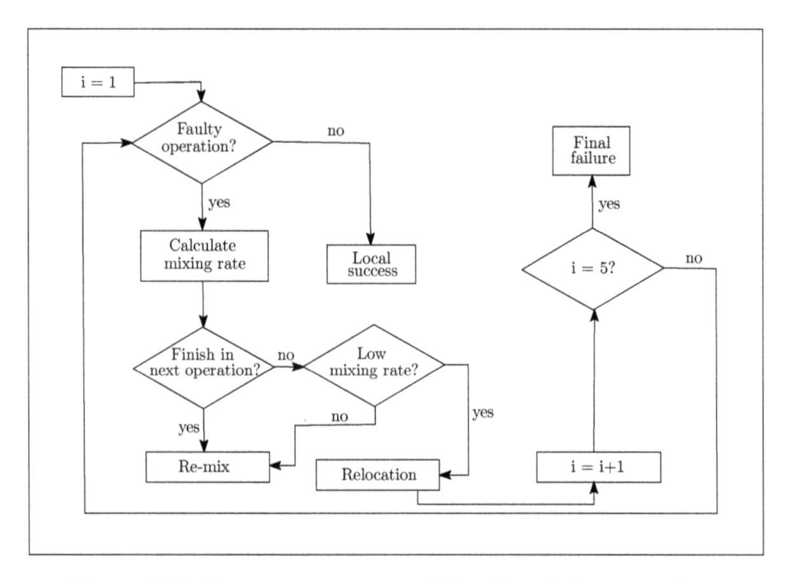

Figure 3.12: Error-recovery control flow for mixing operations

Authors in [193] also proposed a roll-forward approach to reduce error-recovery time. It attempts to perform redundant operations by utilizing the unused regions on the chip. This approach anticipates possible errors and "re-executes extra operations" along with the original flow. It utilizes additional reagent-droplets or available back-up droplets to execute the same operation in two parallel paths. To implement such

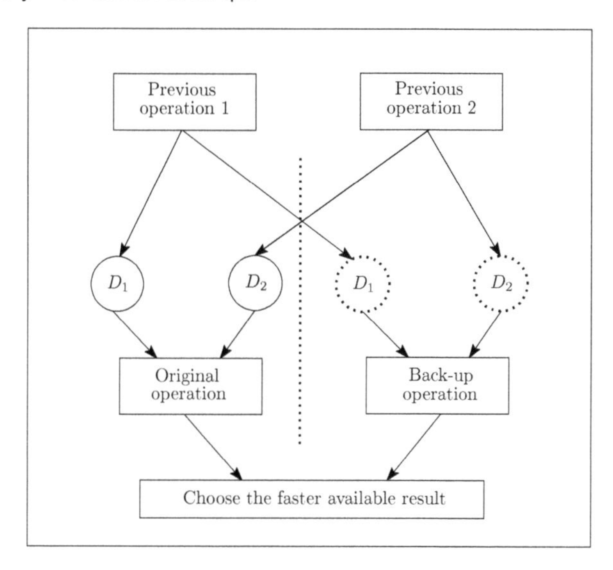

Figure 3.13: Roll-forward error recovery

a redundant operation, two-droplets of each reagent are dispensed at the beginning of the assay. The flow of roll-forward-based error recovery is shown in Fig. 3.13. It consists of the following two steps:

- **Step 1:** First, both *previous operation* 1 and *previous operation* 2 are executed to produce two daughter-droplets, each. One droplet is used in the subsequent operation, and the other one is stored as a back-up/copy-droplet (to carry out the same operation in parallel for error-recovery). The subsequent operation is referred to as *Original operation*. This is carried out using droplets D_1 and D_2. A redundant operation is also performed simultaneously using the back-up droplets corresponding to D_1 and D_2. This is referred to as back-up operation. These back-up droplets are used immediately without storing them.
- **Step 2:** In this step, the outcomes of the original operation and back-up operation are checked with on-chip sensors at the same time and the following action is invoked as required:
    - If one of them quickly generates a correct droplet (i.e., within the acceptable *CF*-range), then the droplet that is produced earlier is selected for the next operation. The other (slower) ongoing operation is stopped immediately, and the droplet is sent to the waste reservoir.
    - When errors occur both in original and back-up paths, the droplet that finishes error recovery faster is selected.
    - The back-up operation is not carried out in the case of limited on-chip space.

3.4 CONCLUSION

In this chapter, we have reviewed various techniques that have been developed for correcting fluidic errors in DMFB and MEDA-based chips during sample preparation. Errors caused by non-homogeneous mixing and unbalanced split operations are usually considered. However, most of these error-recovery approaches suffer from high overhead, long recovery time, and non-determinism in termination, particularly when multiple errors occur.

Section III

Design Automation Methods

4 Error-Correcting Sample Preparation with Cyber-physical DMFBs

Digital (droplet-based) microfluidic technology offers an attractive platform for implementing a wide variety of biochemical laboratory protocols, such as point-of-care diagnosis, DNA analysis, target detection, and drug discovery. A digital microfluidic biochip consists of a patterned array of electrodes on which tiny fluid droplets are manipulated by electrical-actuation sequences to perform various fluidic operations, e.g., dispense, transport, mix, or split. However, because of the inherent uncertainty of fluidic operations, the outcome of biochemical experiments performed on-chip can be erroneous even if the chip is tested *a priori* and deemed to be defect-free.

This chapter first introduces an important error-recoverability problem in the context of sample preparation using DMFBs, and presents a new technique called "Error-Correcting Sample Preparation" (ECSP) [122]. ECSP requires a cyber-physical platform so that physical errors, when detected online at selected checkpoints with integrated sensors, can be corrected by invoking recovery techniques. As discussed in the previous chapter, almost all prior work on error recoverability used checkpointing-based *rollback* approaches, i.e., re-execution of the affected portion of the protocol starting from the previous checkpoint. Unfortunately, such techniques are expensive both in terms of assay-completion time and reagent cost, and are unable to ensure full error-recovery in a deterministic sense. In order to address error recoverability from imprecise droplet mix-split operations, ECSP relies on a novel *roll-forward* approach where the erroneous droplets, instead of being discarded, are used in the error-recovery process. All erroneous droplets participate in the dilution process, and they mutually cancel or reduce the concentration error when the target droplet is reached. A rigorous analysis that reveals the impact of volumetric errors on target-concentration is presented. The layout of an LoC that can execute the proposed roll-forward algorithm is also designed. The analysis reveals that fluidic errors caused by unbalanced droplet splitting can be classified as being either *critical* or *non-critical*, and only those of the former type require correction to achieve error-free sample dilution. Simulation experiments on various sample-preparation test-cases demonstrate the effectiveness of ECSP.

The rest of the chapter is organized as follows: Basic preliminaries of sample preparation, prior work on error-recovery, and the idea behind the roll-forward approach are discussed in Section 4.1. Theoretical results on volumetric-error correction and the motivation behind this work are presented in Section 4.2. A formal statement of the problem is also presented in this section. Details of error-correcting sample preparation are discussed in Section 4.3. The physical layout of a biochip

DOI: 10.1201/9781003219651-4

that can be used to implement ECSP is described in Section 4.4. In Section 4.5, we present experimental details and comparative results with prior rollback-based recovery approaches for a number of real-life as well as synthetic test cases. Finally, conclusions and open problems are drawn in Section 4.6.

4.1 AUTOMATED SAMPLE PREPARATION

4.1.1 RELATED PRIOR WORK

Several algorithms for sample preparation with DMFB have been reported in the literature [100, 113, 131, 136, 163]. A simple static simulator was designed by Maciej *et al.* [103] for visualizing the set of fluidic operations currently being executed on the chip; this simulator also attempts to display the effect of an error on the droplet volume. Another software tool was developed by Grissom *et al.* [49] which aids to analyze sample-preparation algorithms with DMFBs. However, the problem of error-recovery during sample preparation was not studied earlier. Previous work on cyber-physical DMFBs focused on the general error recovery problem in assays, and they are all based on either a rollback and re-execution strategy, or the usage of pre-stored sequence [4,58,59,66,105,106,192]. Zhao *et al.* [192] proposed a re-execution-based method for error-recovery, but it suffers from major drawbacks. During recovery, all ongoing operations of the assay are to be stopped, even those which do not depend on erroneous droplets. Further, the recovery subgraph is assumed to be devoid of any error. In the pre-stored sequence approach, some error recovery sequences are stored in memory before the actual execution of the assay [3, 106]. Although such schemes may reduce the error-recovery time significantly, they can tolerate only a few errors at the cost of a large amount of additional memory. In order to reduce the size of error-recovery subgraphs, Hseih *et al.* [58] and Luo *et al.* [105] suggested re-execution of a portion of the assay starting from some intermediate droplets (also called back-up droplets) stored earlier. However, they require on-chip storage, and impose additional routing constraints on other ongoing assay operations. All the error-recovery approaches described above re-execute the recovery subgraph when an error is sensed, and erroneous droplets are discarded or re-mixed. Recently, Alistar *et al.* [4] proposed a probabilistic solution, where the underlying sequencing graph is converted into a fault-tolerant sequencing graph; each split is attempted a fixed number of times for possible error recovery. However, such repetitive execution not only causes loss of expensive reagents but also increases assay-completion time. Unlike all traditional schemes, we describe below an online roll-forward strategy to achieve volumetric split-error tolerance. The advantages of this technique named as ECSP, are highlighted in Table 4.1 in contrast to prior rollback approaches.

4.1.2 ROLL-FORWARD SCHEME FOR ERROR RECOVERY

Poddar *et al.* [122] introduced a novel *roll-forward* scheme to implement error-correcting sample preparation in a cyber-physical digital microfluidic platform. The main features of this work are summarized below.

Table 4.1
Comparative features of the roll-forward approach against prior art.

Method	Recovery Strategy	Recovery guaranteed?	Erroneous droplets re-used	Reliable for multiple errors	Extra time required?	Extra storage required?	Assay stalls?
[192]	rollback*	no	no	no	yes	yes	yes
[3]	re-merge & re-split**	no	yes (only at source)	no	yes	no	yes
[4]	rollback*	no	no	no	yes	yes	yes
[105]	rollback	no	no	no	yes	no	no
[106]	rollback	no	no	no	yes	yes	no
[58]	rollback (dynamic)	no	no	no	yes	yes	no
[122]	roll-forward (ECSP)	yes	yes	yes	no	no	no

* all operations within the error-recovery subgraph are assumed to be error-free.
** all split operations are allowed to be executed a fixed number of times and the last one is assumed to be error-free.

- We describe an error-correcting sample preparation scheme (ECSP) that automatically corrects the concentration factor of a target in the presence of volumetric split-errors on a cyber-physical digital microfluidic biochip.
- Multiple occurrences of volumetric split-errors on the reaction path are considered and ECSP utilizes a *roll-forward* approach that proceeds with the erroneous droplets to cancel-and-correct concentration error, if any, at the target.
- Volumetric split-errors are classified as being either critical or non-critical, and corrective measures are initiated only for the critical errors on the basis of sensor feedback.
- ECSP provides 100% error recovery in contrast to previous approaches, which may fail to correct the error by deploying "re-merge and re-split" techniques, or by using dynamic recovery that re-executes a portion of the assay.

- Previous methods require a large number of on-chip storage cells for saving error-free back-up droplets, which may block routing paths of other droplets on the chip. In contrast, ECSP does not need to store any back-up droplets.
- Since ECSP does not have to repeat earlier mix-split steps or re-execute a portion of the assay from previous checkpoints, it saves assay-completion time significantly compared to rollback approaches.
- ECSP is easy to implement on a biochip compared to prior approaches.

4.2 MOTIVATION AND PROBLEM FORMULATION

In an ideal scenario, a $2X$-size droplet should be split into two $1X$-size droplets. However, successful splitting depends on various parameters such as channel gap, droplet size, degree of contact angle changed by electrowetting on dielectric (EWOD), and concurrent switching of actuation voltages. A split may be balanced, unbalanced, or imperfect as shown in Fig. 1.6. In balanced (unbalanced) splitting, two equal (un-equal) volume droplets are produced, whereas, during imperfect splitting, apart from two equal/unequal size volume droplets, a small residue may be left behind on the middle electrode [75, 175]. Such imperfect splitting not only causes cross-contamination [189] but also adversely affects the accuracy of assay outcome [129].

Although an ideal mix-split operation mandates equal-size splitting, because of several imprecisions in fluidic operations, actuation timing, or electrical uncertainties, it is very hard to achieve such a perfect split in practice. As a result, every split operation that is performed following mixing or during dispensing is susceptible to volumetric imbalance, and in turn, may affect the correctness of the dilution step. Since sample preparation with digital microfluidics essentially consists of a sequence of mix-split operations, such an error may occur in every step of the reaction path. Thus, from the viewpoint of error management, it is desirable to choose those algorithms that attempt to minimize the number of mix-split operations during sample preparation.

To the best of our knowledge, *twoWayMix* (*TWM*) algorithm proposed by Thies *et al.* [163] requires the minimum number of (1:1) mix-split operations for producing a desired target concentration factor with certain accuracy. One of the interesting aspects of *TWM* algorithm is that for a given accuracy level n, the set of target-*CF*s can be partitioned into n disjoint groups based on the number of mix-split steps. For example, the number of mix-split operations required for generating any odd target-*CF* is equal to the accuracy level n; as an instance, the *TWM* algorithm takes nine mix-split operations for generating $\frac{241}{512}$ $(011110001)_2$, because the least significant bit of the *CF* is '1'. In general, the number of target-*CF*s in a group requiring n' operations is determined by the following expression:

$$2^{n'-1}, n' = 1, 2, \ldots, n \tag{4.1}$$

Fig. 4.1 shows a histogram that indicates the number of target-*CF*s in different groups, for accuracy $n = 9$, where each group needs the same number of mix-split operations.

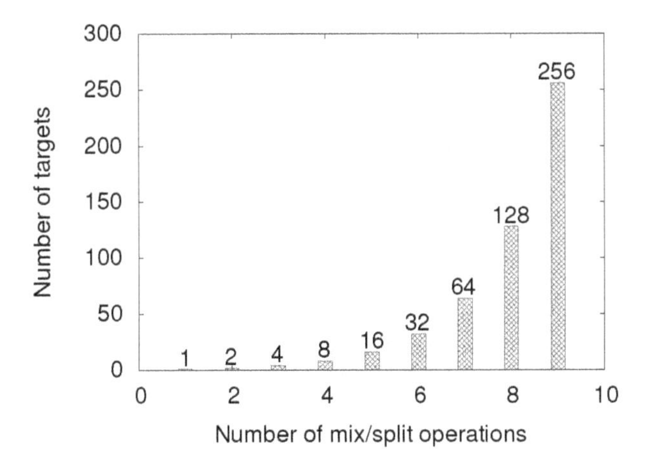

Figure 4.1: Total number targets in different groups for accuracy $n = 9$ (with permission from ACM [122])

4.2.1 ERROR MODELING: EFFECT OF ERRORS ON TARGET-CFS

It has been observed that erroneous sample preparation is caused by variations in droplet volumes in more than 80% of the instances [129]. An intriguing question in this direction, is the following:

"How is the concentration factor of the target droplet affected when a volumetric error ε occurs during the execution of the reaction path?".

In order to answer this question, we consider the following error model. We assume that a dispense operation from a reservoir is error-free. A dispense operation that causes a volumetric error in droplet size can be handled by returning the droplet to the source reservoir when detected by a sensor, and re-dispensing it. Such operations do not waste any reactant and do not take much time as there are only a few operations to be re-executed. Other techniques based on droplet-emission dynamics to correct dispense errors are also known [27,46]. On the other hand, when an unbalanced split operation occurs, correction becomes expensive, and such an error needs careful analysis. We assume that the volumes of the two resulting droplets become (1 + ε) and (1 - ε), where $0 \leq \varepsilon < 1$ denotes the volumetric error [10]. Such volumetric variation in an intermediate step may cause concentration errors in target droplets. Typically, $\varepsilon \leq .07$ for a split operation with DMFBs [129]. For example, consider the dilution problem shown in Fig. 4.2, where S_i represents the ith (1:1) mix-split operation in the sequence, C_i the resulting concentration factor, r_i the *CF* of sample or buffer added at the ith step, and C_t represents the desired target-*CF*. Let us assume that an unbalanced-split volumetric split-error ε occurs after the first operation (S_1) on the reaction path.

As a result of this volumetric error at S_1, the volume of a resultant droplet becomes ($1 \pm\varepsilon$). This error will cause a concentration error in the next mix-split operation, and as a result, the concentration and volume of the resulting droplets after

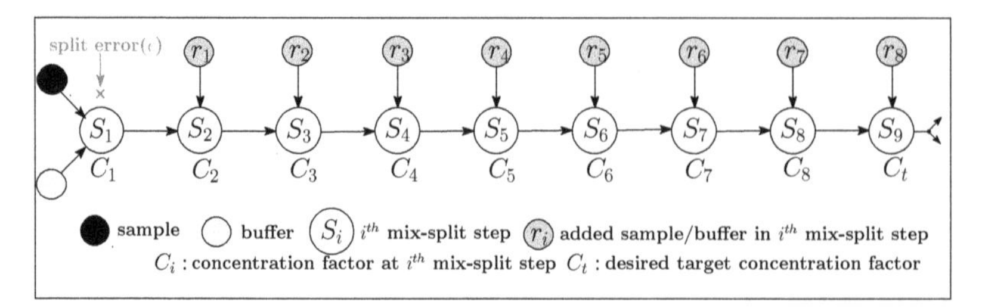

Figure 4.2: Mix-split operations for generating target concentration C_t with accuracy level $n = 9$ (with permission from ACM [122])

operation S_2 will become: $C_2 = \frac{(C_1)(1\pm\varepsilon)+r_1}{1+(1\pm\varepsilon)}$, and $V_2 = \frac{2\pm\varepsilon}{2}$, where $r_i = 0(1)$ if sample (buffer) is added in this step. The sign $(+/-)$ in these expressions is chosen depending on whether the larger or the smaller droplet is used at the next step following a split. An erroneous droplet, when mixed with the sample/buffer droplet in the next operation, affects the concentration and volume of product droplets. As a result, at the end of operation S_3, the concentration and volume of the droplet would become: $C_3 = \frac{C_1(1\pm\varepsilon)+r_1+2r_2}{[2^2\pm\varepsilon]}$, and $V_3 = \frac{(2^2\pm\varepsilon)}{2^2}$, respectively. In this way, an error originated in some step, continues to propagate along the reaction path and may alter the concentration factor of the target droplet at the end. For the above example, when a volumetric error occurs in Step 1, the erroneous *CF* and volume of the target droplet will be given by:

$$C^1_{error} = \frac{C_1(1\pm\varepsilon)+r_1+2r_2+2^2r_3+2^3r_4+2^4r_5+2^5r_6+2^6r_7+2^7r_8}{[2^8\pm\varepsilon]} \tag{4.2}$$

$$V^1_{error} = \frac{(2^8\pm\varepsilon)}{2^8} \tag{4.3}$$

Note that in the absence of any error on the reaction path, the concentration factor and volume of the target droplet are given by:

$$C^1_{correct} = \frac{C_1+r_1+2r_2+2^2r_3+2^3r_4+2^4r_5+2^5r_6+2^6r_7+2^7r_8}{2^8} \tag{4.4}$$

$$V^1_{correct} = \frac{2^8}{2^8} = 1 \tag{4.5}$$

However, if a volumetric split-error occurs during mix-split operation S_2, the expressions for erroneous target-*CF* and volume will be:

$$C^2_{error} = \frac{(C_1+r_1)(1\pm\varepsilon)+2b_r+2^2r_3+2^3r_4+2^4r_5+2^5r_6+2^6r_7+2^7r_8}{2[2^7\pm\varepsilon]} \tag{4.6}$$

$$V^2_{error} = \frac{(2^7\pm\varepsilon)}{2^7} \tag{4.7}$$

Thus, the impact of a volumetric split-error occurring at a mix-split step, on a target-*CF*, can be precomputed using the above expressions. Necessary cyber-physical action may be taken when the volumetric split-error affects the concentration of the target badly (i.e., when the error in target-*CF* exceeds certain threshold).

4.2.2 IMPACT OF MULTIPLE ERRORS ON TARGET-CF: ERROR COLLAPSING

Since the dilution process consists of a sequence of n mix-split operations, one or more split-errors may occur in any step along the reaction path. We now show that a multiple split-error can be considered to be equivalent to a single split-error located further down on the path, as far as its impact on the target-*CF* is concerned.

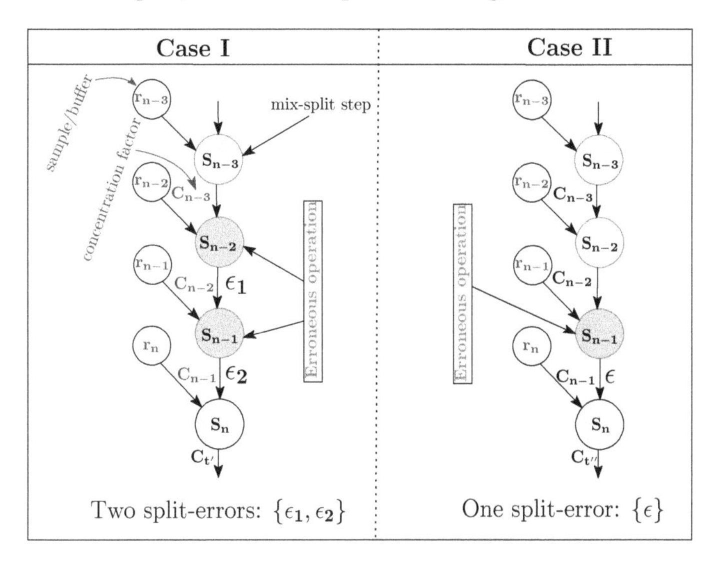

Figure 4.3: Volumetric error occurring during different operations on the reaction path (with permission from ACM [122])

Definition 1 Two volumetric split-errors ε_1 and ε_2 are said to be *equivalent* if the generated target-*CF* (approximated as a rounded n-bit binary fraction for a given accuracy level n) in the presence of error ε_1, is the same as that produced by error ε_2.

Example 1 Let us explain the concept with the example given in Fig. 4.3. As depicted in Case I of Fig. 4.3, a multiple volumetric error $\{\varepsilon_1, \varepsilon_2\}$ occurs at steps S_{n-2} and S_{n-1} on the reaction path. As a result of the split-error at S_{n-2}, the *CF* and volume of the resulting droplets after operation S_{n-1} become: $C_{n-1} = \frac{C_{n-2}(1+\varepsilon_1)+r_{n-1}}{2+\varepsilon_1}$, and $V_{n-1} = \frac{(2+\varepsilon_1)}{2}$; where $C_{n-2} = \frac{(C_{n-3}+r_{n-2})}{2}$. Next, when error ε_2 occurs at S_{n-1}, the *CF* and volume of the outcome droplets at operation S_n are given by: $C_{t'} = \frac{C_{n-1}(V_{n-1}+\varepsilon_2)+r_n}{V_{n-1}+1+\varepsilon_2}$, and $V_{t'} = \frac{(V_{n-1}+1+\varepsilon_2)}{2}$.

Now consider a single error ε that occurs at operation S_{n-1}, shown as Case II in Fig. 4.3. The CF and volume of the droplets after operation S_{n-1} will be $C_{n-1} = \frac{(C_{n-2}+r_{n-1})}{2}$ and $V_{n-1} = 1$, where $C_{n-2} = \frac{(C_{n-3}+r_{n-2})}{2}$. Finally, the CF and volume of the target droplets can be expressed as: $C_{t''} = \frac{C_{n-1}(1+\varepsilon)+r_n}{2+\varepsilon}$ and $V_{t''} = \frac{(2+\varepsilon)}{2}$. The multiple error $\{\varepsilon_1, \varepsilon_2\}$ is said to be equivalent to a single error ε if $C_{t'} = C_{t''}$. By equating the expressions for $C_{t'}$ and $C_{t''}$, we can solve for ε in terms of ε_1, ε_2, and other known parameters to determine the equivalent single error at step $S_{(n-1)}$. This analysis leads to the following result.

Lemma 1 It is always possible to collapse multiple split-errors to an equivalent single split-error whose effect on the final concentration error of the target droplet is same as that of the multiple errors.

Table 4.2
Equivalence of errors for different target-*CF*s and accuracy n = 9.

Target-CF	$P(\varepsilon_1, \varepsilon_2)$	$\{\varepsilon_1, \varepsilon_2\}$	$P(\varepsilon)$	$\{\varepsilon\}$	% error in target-CF
$\frac{23}{512}$	{7,8}	{3%,2%}	{8}	6.6%	3.17%
$\frac{51}{512}$	{7,8}	{4%,−3%}	{8}	2.9%	1.41%
$\frac{183}{512}$	{7,8}	{−2%,6%}	{8}	5.81%	2.82%
$\frac{249}{512}$	{7,8}	{4%,2%}	{8}	3.92%	1.91%

The result stated in the above lemma helps us to reduce the size of simulation experiments considerably. Table 4.2 shows the single-error equivalence of a double-error for accuracy level $n = 9$. Column 1 represents the target-CFs, Column 2 shows the locations (i.e., step numbers) where double-error has occurred; Column 3 shows the corresponding percentage volumetric error. The fourth and fifth columns denote the location and the percentage of volumetric error for an equivalent single error, respectively. The last column shows the error in target-CF for both cases.

4.2.3 CRITICAL AND NON-CRITICAL ERRORS

As we have observed, an unbalanced-split-error on any step of the reaction path will change the desired target-CF, thereby causing an error. It has been reported that the maximum volume variation that can occur during the split operation is 7% [129]. However, some of them may be very insignificant to be perceived at the end of the reaction path depending on the magnitude and the position where the error has occurred. Note that there is also an inherent limitation on the accuracy of target-CF, which is determined by the value of n, i.e., how many significant bits have been used in approximating the target concentration. For example, when $n = 9$, at most nine mix-split steps are needed by the *TWM* method [163], and for such a choice of n, an accuracy of more than $\frac{1}{1024}$ cannot be achieved for a target-*CF*. Clearly, there is a trade-off between accuracy, and dilution time and cost.

Definition 2 If the error of a target-*CF* induced by an unbalanced-split is less than the inherent accuracy level that is used in the dilution algorithm, then it can easily be ignored. Clearly, there will be no perceptible change at the output *CF*, which is approximated as an n-bit fractional binary number. We call such errors as *non-critical*, and the rest as *critical*. Clearly, only critical errors visibly affect the target-*CF*, and when they are detected by sensors, corrective measures are needed to restore the desired concentration factor of the diluted sample.

Table 4.3
Non-critical split errors for *CF* = $\frac{241}{512}$ and accuracy n = 9.

Error position	Produced *CF*	Concentration change (%)	Non-critical?
1	$\frac{241.004}{512}$	+0.001%	Yes
2	$\frac{240.938}{512}$	-0.025%	Yes
3	$\frac{240.806}{512}$	-0.080%	Yes
4	$\frac{240.543}{512}$	-0.189%	Yes
5	$\frac{241.135}{512}$	+0.056%	Yes
6	$\frac{242.309}{512}$	+0.543%	No
7	$\frac{244.628}{512}$	+1.505%	No
8	$\frac{249.149}{512}$	+3.380%	No

Example 2 In order to understand the above concept, let us illustrate it with an example. We assume that $n = 9$, and let us consider target-*CF* = $\frac{241}{512}$. Also assume that an unbalanced-split-error has occurred at Step S_1, and that the amount of volumetric error (ε) is 7%. As a result of this volumetric imbalance, the error in target-*CF* will be given by: $E_{0.07} = \frac{0.5(1\pm0.07)+2^3+2^4+2^5+2^6}{[2^8\pm0.07]}$ - $\frac{241}{512} = 0.000008 < \frac{1}{1024}$. Hence, the error in target-*CF* is less than $\frac{1}{2^{n+1}}$, the inherent accuracy of the dilution algorithm. Clearly, the effect of such a volumetric split-error on the target-*CF* cannot be perceived within the given approximation limit, and hence this split-error is *non-critical*. If the error in target-*CF* becomes greater than or equal to $\frac{1}{2^{n+1}}$, the corresponding split-error is named *critical*. Our simulation experiments reveal that there are many non-critical split-errors on a reaction path for a given target-*CF*. Table 4.3 lists the criticality of operations at different mix-split steps for the target-*CF* = $\frac{241}{512}$, assuming that a volumetric split-error of 7% has occurred.

Needless to say, for a given target-*CF*, the number of critical split operations tends to increase when the amount of volumetric error increases. Fig. 4.4 shows the

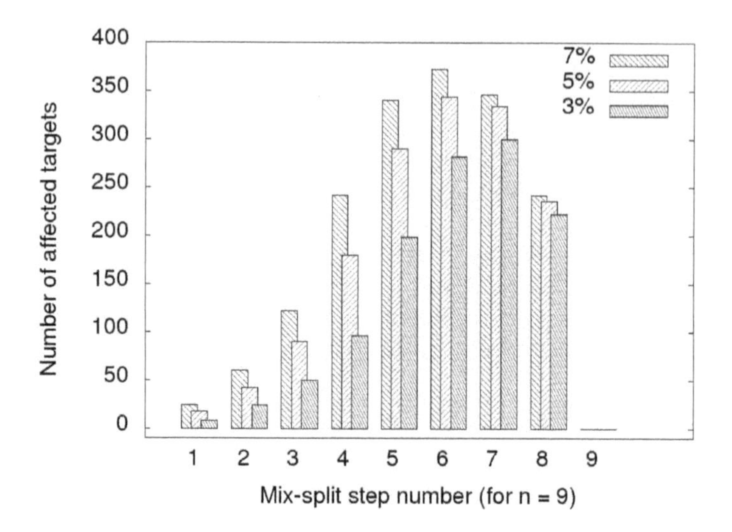

Figure 4.4: Number of targets affected by a critical error occurring in a particular mix-split step (with permission from ACM [122])

number of target-*CF*s that are affected when a critical split-error ε occurs at a given mix-split step. In our experiment, we have chosen $n = 9$ and $\varepsilon = 3\%$, 5%, and 7%.

Fig. 4.4 also reveals that a small number of target-*CF*s are affected during early mix-split operations of the sequence. Their number increases gradually until the middle range and again drops further down along the mix-split sequence. Note that the last mix-split operation (for example, the 9th step) cannot be critical because a volumetric error therein only imbalances the size of two target droplets and cannot change the *CF*. In addition, it has been observed experimentally that for each group of target-*CF*s as shown in Fig. 4.1, there may be some specific mix-split steps which are non-critical. For example, consider the group of target-*CF*s that need all nine mix-split steps (i.e., for all *CF*s = $\frac{x}{2^9}$, where x is an odd integer such that $1 \leq x \leq 511$). The Steps $\{1, 2, 3\}$ are observed to be non-critical for all of them when $\varepsilon = 7\%$. Furthermore, Fig. 4.5 shows that for accuracy level $n = 9$, none of the target-*CF*s will have critical errors of multiplicity[1] 6, 7, 8 or 9. Thus, for achieving error-tolerance, a small number of checking is required.

Fig. 4.6 depicts an interesting error behavior that is observed while diluting a sample. The red solid curve shows the plot of non-perceptible errors in target-*CF* that is determined by the accuracy (approximation) limit of the algorithm, against various *CF* values. The pink dashed (black dotted) curve shows the errors in target-*CF* when a volumetric split-error of 5% (7%) is inserted at the last-but-one-step of the reaction path, against various *CF*s. Recall that all multiple split-errors can be collapsed to single split-error at the end of the reaction path, and hence, an error at this particular step is sufficient to capture all effects of multiple errors occurring elsewhere. It is also observed that the maximum error in target-*CF* is 3.38% for 7%

[1]total number of critical errors for a particular target-*CF*.

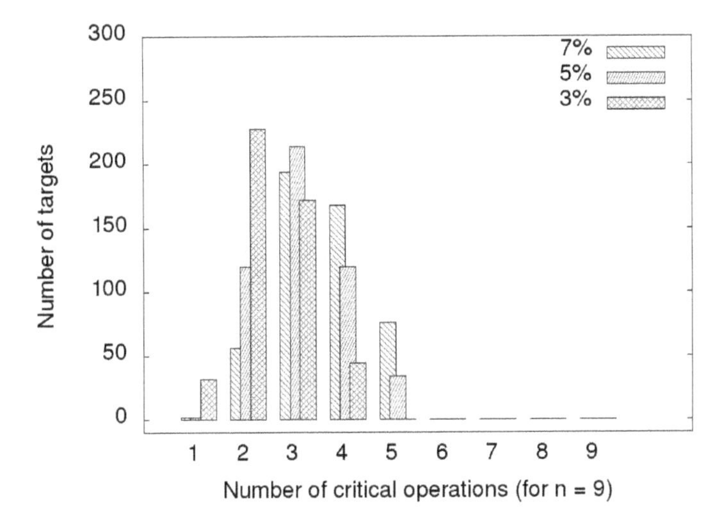

Figure 4.5: Number of targets affected by critical split-errors of different multiplicities (with permission from ACM [122])

volumetric error, when $n = 9$. Also, these two curves show that the concentration error remains constant for all *CF* values lying within $\frac{1}{2^n}$ to $\frac{1}{2}$. The reason behind this is that for all target-*CF*s up to $\frac{1}{2}$, the least significant bit in each of their bit representations is 0, and hence, the percentage-error remains constant for these target-*CF*s. Also, there is a sharp dip to zero in concentration error when the target-*CF* is exactly $\frac{1}{2}$; because only one mix-split is needed to generate this target concentration.

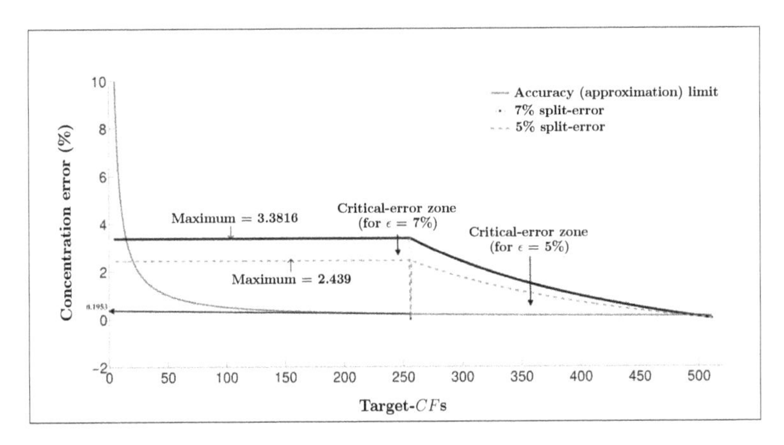

Figure 4.6: Concentration error (%) for different *CF*s (with permission from ACM [122])

Note that any concentration error that lies below the red solid curve is not perceptible because of the approximation limit, whereas any error that lies above the black dotted or pink dashed curve will not arise when the magnitude of volumetric error is bounded above by the given ε. Hence, the critical region where error correction is needed, corresponds to the region that is bounded between these two curves. In practice, the number of critical split-errors may be even less because for some of them, the error in target-*CF* may be too small compared to sensor sensitivity. On the basis of this observation, critical split-errors can easily be enumerated *a priori* by simulation, given a target-*CF* and accuracy level. Hence, only for those mix-split steps, the resulting droplets are to be tested for volumetric errors, if any, by transporting them to on-chip sensor locations. Appropriate corrective procedures may also be executed accordingly when the sensor feedback is positive.

4.2.4 CANCELLATION OF CONCENTRATION ERROR AT THE TARGET

When an unbalanced-split-error occurs following (1:1) mixing at some step of the reaction path, two droplets, one having larger volume (1 + ε) and the other with smaller volume (1 - ε) are produced. In the *TWM* dilution algorithm [163], only one of them is used in the next step while the other one is discarded. In prior error-recovery techniques for bioassays that deploy cyber-physical correction mechanism [4, 58, 105, 192], when such volumetric imbalance is detected by sensors, either a "re-mix and re-split" strategy is adopted to correct the error, or the assay is re-executed from a previous checkpoint using good left-over droplets that were produced or stored earlier. Furthermore, no distinction was previously made between critical and non-critical errors, and hence, unnecessary corrective measures had to be taken even for non-critical errors.

The ECSP method of error-correction adopts a fundamentally different strategy. We do not *undo* an erroneous fluidic operation. We observe that when a volumetric error occurs following a split operation, the two daughter-droplets, thus obtained, will have errors with equal magnitude but with opposite signs. Hence, when a critical error occurs, instead of redoing an erroneous split, we proceed with a duplicate reaction path with the discarded sibling in addition to the original path in the dilution assay, and mix together an appropriate droplet-pair produced thereafter. As a result, the concentration error propagated along the two paths mutually cancel each other when the target is reached. We now formally present some theoretical results.

Consider the dilution problem of generating a desired target concentration $\frac{c_i}{2^n}$ with accuracy level n. Let $\{\ldots S_{i-1}, S_i, S_{i+1}, S_{i+2}, S_{i+3}\}$ represent the (1:1) mix-split sequence that produces the target-*CF* when the *TWM* algorithm [163] is run on a digital microfluidic device. Assume that a volumetric split-error of magnitude ε occurs at operation S_{i+1} producing two droplets of volume (1 + ε) and (1 - ε). Fig. 4.7 depicts the mix-split steps where the leftmost path corresponds to the regular reaction sequence. The right sub-tree of node S_{i+1} will be executed only if the split operation at Step S_{i+1} is critical (known *a-priori*) and the sensor feedback is positive (known online). Let $T(S_{i+1})$ denote the sub-tree of the complete mix-split tree rooted at node S_{i+1}.

Let d_j, d'_j, ..., d_{j+2}, d'_{j+2}, ...,d_{j+3}, d'_{j+3} denote the target droplets as shown in Fig. 4.7. Note that they appear as leaf nodes in the mix-split tree, and in ideal condition (when no split-error occurs anywhere) each of them will have the same value of CF (the desired target concentration $\frac{C_i}{2^n}$). We can now prove the following result.

Theorem 1 If droplet d_j (a target droplet that appears as a leaf of the left sub-tree of $T(S_{i+1})$) and droplet d_{j+2} (a target droplet that appears as a leaf of the right sub-tree of $T(S_{i+1})$) are mixed together and then split, two droplets of $CF = \frac{C_i}{2^n}$ will be correctly produced even if the mix-split step occurring at their nearest-common-ancestor node (i.e., Step S_{i+1} of the mix-split tree) is affected by a volumetric split-error (ε).

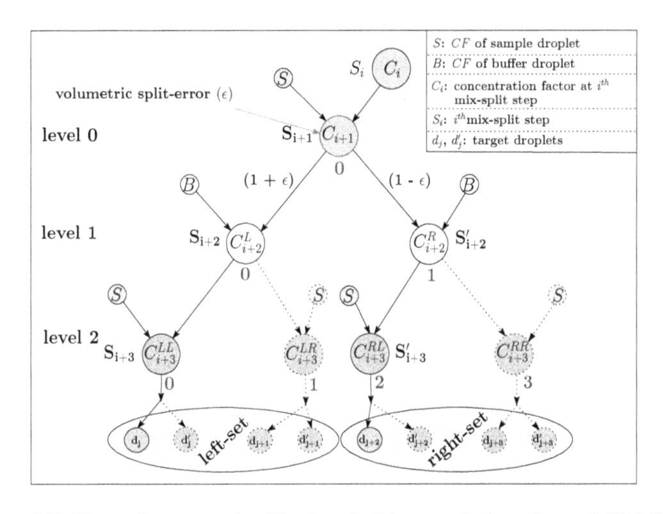

Figure 4.7: Error-free sample dilution (with permission from ACM [122])

Proof In order to prove the claim, we show that the target concentration observed in the presence of such an error is same as the desired value in the absence of any error. As shown in Fig. 4.7, the desired value of $CF = \frac{C_{i+1}+2*S}{4}$, where $C_{i+1} = \frac{(C_i+S)}{2}$. Note that in the absence of an error, the CF values of all droplets lying in the same level of the tree will be equal. When an unbalanced split-error (ε) occurs at Step $\{S_{i+1}\}$, two droplets of volume $(1+\varepsilon)$ and $(1-\varepsilon)$ are produced. Without any loss of generality, we assume that a buffer droplet is mixed next; if both siblings are now used in two identical reaction paths, the left sub-tree and right sub-tree will produce droplets with the following CFs and volumes (the superscripts L and R denoting left and right components): $C^L_{i+2} = \frac{C_{i+1}*(1+\varepsilon)}{1+(1+\varepsilon)}$, $C^R_{i+2} = \frac{C_{i+1}*(1-\varepsilon)}{1+(1-\varepsilon)}$, $V^L_{i+2} = \frac{1+(1+\varepsilon)}{2}$, and $V^R_{i+2} = \frac{1+(1-\varepsilon)}{2}$. Thus, two pairs of droplets, each with concentration factor C^L_{i+2} and C^R_{i+2} are generated. Finally, let us assume that a sample droplet is used in the last step along the left arm of each subtree; as a result, two groups, each consisting of two droplets are produced with CF values given by: $C^{LL}_{i+3} = \frac{C_{i+1}*(1+\varepsilon)+2*S}{1+2+(1+\varepsilon)}$, $C^{RL}_{i+3} = \frac{C_{i+1}*(1-\varepsilon)+2*S}{1+2+(1-\varepsilon)}$, and with the volume $V^{LL}_{i+3} = \frac{1+2+(1+\varepsilon)}{4}$ and $V^{RL}_{i+3} = \frac{1+2+(1-\varepsilon)}{4}$, respectively. Let these droplets produced at Steps $\{S_{i+3}\}$ and $\{S'_{i+3}\}$ be named as $(d_j, d'_j, d_{j+2}, d'_{j+2})$ as shown in Fig. 4.7. Note that when droplets d_j and d_{j+2} are mixed together, then the concentration factor of the mixed droplets becomes $\frac{C_{i+1}+2*S}{4}$, which is same as that of the desired target droplet. Similarly, when the pair of droplets (d'_j, d'_{j+2}) are mixed together, the errors carried along the two paths cancel each other, and the target-*CF* is reached correctly. Therefore, by mixing one droplet from the left-set and another one from the right-set (appearing as leaves of two subtrees in Fig. 4.7), the effect of the volumetric error that has occurred at the common root node S_{i+1} can be canceled on reaching the target. This result can be generalized for any sequence of mix-split steps. Hence the proof.

Corollary 1 Let the desired target concentration be $\frac{c_i}{2^n}$ for accuracy level n, and also let $\{S_1, S_2, \ldots, S_{k-2}, S_{k-1}, S_k\}$ be the sequence of mix-split operations for generating the target-*CF*. Let there be a multiple split-error that affects several mix-split steps in a dilution sequence of length k. If all 2^k droplets that are produced by the tree without discarding any waste droplet are mixed together at the end, the *CF* of the resulting mixture will be exactly equal to the desired value $\frac{c_i}{2^n}$ regardless of the number of split-errors. In other words, all errors will be mutually canceled, and further, no checkpoint or detection mechanism is required in this scenario.

Corollary 2 The error in *CF* of a target C_t produced by a volumetric split-error ε occurring at a particular step S_i of the mix-split sequence is exactly the same as that in *CF* of the target (1 - C_t) when the same volumetric split-error ε occurs at the step S_i of a similar mix-split sequence where sample droplets are replaced by buffer droplets and vice-versa. For the sake of better understanding, this fact is graphically depicted in Fig. A.6 (as described in Section A.3 of the Appendix) for some representative target-*CF*s when the accuracy level $n = 9$.

Observation 1 For accuracy level up to $n = 4$ in target-*CF*, we observe some special properties: given the mix-split sequence of any target-*CF* obtained by the *TWM* algorithm, no mix-split step is critical for split-error $\varepsilon \leq 7\%$. In other words, no checkpoint or sensor detection is needed for error correction. Also, for $n = 5$, no target-*CF*s except $\frac{15}{32}$ and $\frac{17}{32}$ are susceptible to volumetric error $\varepsilon \leq 7\%$. Only for these two concentration factors, an error of only 3% may occur when the Step 4 of the mix-split step suffers from a volumetric error. Hence, for ensuring correctness, only one checkpoint will be sufficient.

The theoretical results stated above constitute the basis for the design of the following error-correcting dilution algorithm.

4.2.5 PROBLEM FORMULATION

The objective of the error-correcting dilution scheme (ECSP) is to ensure the correctness of target concentration factor C_t $(0 < C_t < 1)$ under the (1:1) mixing model in the presence of one or more volumetric split-errors. The problem is formally described as follows:

- **Inputs:**
  - a supply of sample fluid ($CF = 1$) and a buffer fluid ($CF = 0$);
  - target concentration factor as C_t, $0 < C_t < 1$;
  - the magnitude of volumetric split-error ε, $0 < \varepsilon < 1$; for evaluation purposes in this chapter, we have assumed $\varepsilon \leq 0.07$;
  - the value of n denoting the accuracy (approximation) of CF;
    Note that C_t is approximated (rounded-off) as $\frac{x_i}{2^n}$, where $x_i \in \mathbb{Z}^+$, $0 < x_i < 2^n$, and $n \in \mathbb{Z}^+$ and n is the accuracy level.
- **Output:**
  - The (1:1) mix-split sequence that will produce C_t within the approximation limit of $\frac{1}{2^{n+1}}$ in the presence of volumetric split-errors, if any, on the basis of sensor feedback.

4.3 ERROR-CORRECTING DILUTION ALGORITHM

Given the input information as stated in 4.2.5, the roll-forward error-correcting scheme (ECSP) for sample preparation, is governed by cyber-physical control online, and it requires the following actions:

- determine the reaction path using the *TWM* algorithm [163], and identify critical mix-split operations as described in Section 4.2.3.
- for critical mix-split steps,
  - (i) an additional actuation sequence is inserted in the mixing algorithm that transports the split droplets to on-chip sensor locations;
  - (ii) a roll-forward actuation sequence is inserted that mimics the original reaction path following an erroneous split;
  - (iii) the original mix-split sequence is modified at the end of the reaction path to cancel the error at the target as described in Section 4.2.4;
- Necessary control sequences are added depending on whether the sensor feedback is positive (i.e., indicating a volumetric split-error online), to activate the roll-forward paths as needed.

4.3.1 REACTION PATH: CRITICAL OPERATION

Algorithm 1, described below, takes the desired target concentration factor C_t, maximum split-error ε and accuracy level n as inputs, and returns the list of critical operations CR_{ops} for the given target concentration factor as output. Initially, lower order zero bits are removed from the binary representation (BS) of C_t until the least significant bit becomes 1, and then the bit-string is reversed. In the *TWM* algorithm [163], these bits will determine whether a sample (if the bit is 1) or buffer (if the bit is 0) is to be mixed with the previously produced droplet at a particular step. We now determine the criticality of a mix-split step by simulation: we insert a volumetric split-error ε for every split operation on the reaction path and check whether or not

Algorithm 1: Critical operations

```
Input: Desired target concentration C_t, maximum split-error ε and accuracy n
Output: List of critical operations CR_ops = {c_r1, c_r2, ... c_rk}
1  Express C_t as an n-bit binary number (BS);
2  Remove the low-order zero bits from BS;
3  Find the length l of BS;
4  Reverse the bits of BS;
5  set CR_ops = Φ;
6  for (i = 1; i ≤ l; i = i + 1) do
7      //calculate the target concentration factor by inserting a volumetric split-error in every step S_i;
8      C'_t = [.5 + Σ_{j=1}^{(i-1)} 2^(j-1) * BS(j)](1±ε) + Σ_{j=i}^{(n-1)} 2^(j-1) * BS(j) / (2^(i-1) * [2^(n-i) ± ε]);
9      E_ε = |(C'_t - C_t)| / 2^n;
10     if E_ε => 1/2^(n+1) then
11         CR_ops ← CR_ops ∪ S_i;
12 return CR_ops
```

$$C'_t = \frac{[.5+\sum_{j=1}^{(i-1)} 2^{(j-1)} * BS(j)](1 \pm \varepsilon) + \sum_{j=i}^{(n-1)} 2^{(j-1)} * BS(j)}{2^{(i-1)} * [2^{(n-i)} \pm \varepsilon]}; \quad E_\varepsilon = \frac{|(C'_t - C_t)|}{2^n}$$

the error in target concentration factor (C'_t) is less than the accepted range $\frac{1}{2^{n+1}}$. In the former case, the volumetric error in the corresponding split operation is marked as non-critical, otherwise, it is flagged as critical. Finally, the list of all critical operations for the given target concentration factor is returned by Algorithm 1.

As for example, suppose the target concentration C_t to be generated is $\frac{241}{512}$ for accuracy $n = 9$. The binary representation (BS) of C_t within this approximation limit is 011110001, and hence, the number of mix-split operations needed to produce C_t will be nine when *TWM* algorithm is run. Now, the list of critical operations (CR_{ops}) returned by Algorithm 1 is {6,7,8} for 7% error (ε). Hence, checkpoints are to be inserted following Steps {6,7,8} for target-error correction. The underlying principle of roll-forward error recovery is described in Algorithm 2 below.

4.3.2 ROLL-FORWARD ERROR RECOVERY

In order to measure the volume of a split droplet, various capacitative or optical sensors/detectors can be used, which are integrated on-chip with the microfluidic device [4, 16, 36, 59, 105, 126]. A capacitive sensing circuit can operate at a high frequency relative to microfluidic droplet velocity [126], hence, detection time can be ignored with respect to assay time.

The key idea behind the proposed error-correcting scheme is to correct any error, which might have occurred during a split operation, when the target is reached. In order to implement this, we perform the remaining mix-split operations following two identical parallel paths (called mirror-paths) in the mix-split task graph starting from the step where a critical error is detected by a sensor. The two erroneous droplets of volume (1 + ε) and (1 - ε) produced therein are used along these two paths, one droplet for each path. These two erroneous droplets produced due to an unbalanced split have equal but opposite volumetric error compared to the ideal case. Hence, they will produce an opposite (in sign) concentration error if the target droplet is prepared using each of them separately following the same reaction path, provided no other error occurs thereafter. Thus, the desired target concentration can be re-generated at

Algorithm 2: Roll-forward error recovery

```
  Input: target-CF: C_t, split-error: ε, accuracy: n
  Output: Desired target-CF within the error limit < 1/2^(n+1)
1 Approximate C_t as x_t/2^n, Flag = 0;
2 Let S = {s_1, s_2, ... s_n} be the sequence of mix-split steps;
3 Let CR_ops = {cr_1, cr_2, ... cr_k} be the list of critical operations returned by Algorithm 1;
4 for (i = 0; i < |S| ; i++) do
5     if S[i] ∉ CR_ops then
6         continue;
7     else
8         if split-error (ε) == 0 then
9             continue;
10        else
11            if Flag == 0 then
12                (i) Let S^1 = S[i] and Label(S[i]) = 0, Flag = 1;
13                (ii) Perform remaining operations along another parallel path using
14                erroneous droplets produced at S^1;
15            else
16                //for multiple errors
17                (i) Let Mirror(S[i]) = S^2 with respect to S^1.;
18                (ii) Perform remaining operations from S[i] and S^2 using the
19                erroneous droplets produced therein.
```

the end of the reaction path by mixing the two droplets at the end. In this context, we introduce the following notation.

When a mix-split step becomes erroneous, we expand the original reaction path into a tree called *error tree*, where the erroneous step serves as the root node (level 0) of the tree (see Fig. 4.7). With each node m of the error tree, we associate two numbers, namely *level number* $L(m)$ and *index number* $I(m)$, where $L(m)$ denotes the level of node m in the error tree, and $I(m)$ denotes the index number of node m in that level (index numbers are assigned from left-to-right starting with 0 in each level). Let $T(L(m))$ denote the total number of nodes in level $L(m)$. We now define the mirror node $M(m)$ of m as the node whose level number is $L(m)$, and index number is $\{I(m) + 2^{L(m)-1}\}$ *mod* $T(L(m))$. For example, the mirror node of node S_{i+2} (level number = 1, index number = 0) with respect to root node S_{i+1}, will have an index number $= \{0 + 2^{1-1}\}$ *mod* $4 = 1$, which is nothing but node S'_{i+2} in Fig. 4.7 (index numbers are shown in blue color). Similarly, the mirror of node S_{i+3} will be S'_{i+3} in level 2.

Algorithm 2 takes the target-*CF*, maximum split-error ε, and accuracy n as input. Initially, checkpoints are inserted on critical steps, the list of which (CR_{ops}) is returned by Algorithm 1. During the execution of a biochemical assay on a DMFB, on-chip sensors/detectors send timely feedback following a possible critical split operation to the micro-controller. Depending on the feedback received from the sensors, the controller performs the following actions:

Case I. If no volumetric error is observed by the sensors in a critical step, then start the next operation as usual; no additional action is needed.

Case II. A volumetric split-error at a critical step is detected by the sensors; consider the error tree rooted at this erroneous node. The two erroneous droplets of

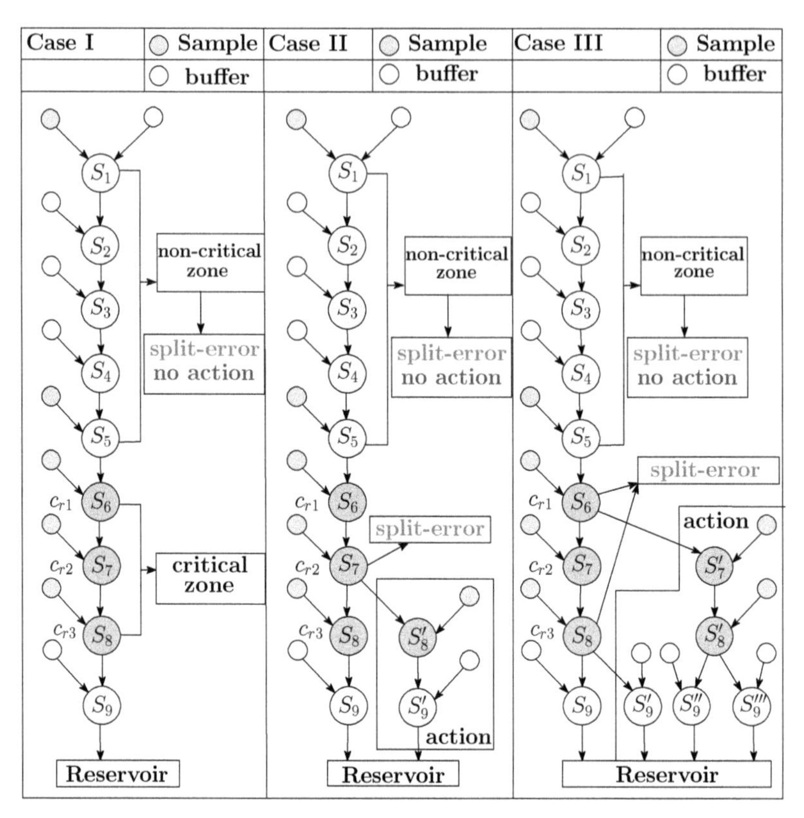

Figure 4.8: Proposed error-correcting dilution scheme (ECSP) for target-$CF = \frac{241}{512}$ (with permission from ACM [122])

volume $(1 + \varepsilon)$ and $(1 - \varepsilon)$ are now available. Perform the remaining mix-split operations following two parallel paths, one along the original one, and the other through the corresponding mirror nodes until the end of the reaction path. If no further critical error occurs, the droplets available at the end are mixed together in a reservoir, as shown in Fig. 4.8.

Case III. If more than one errors are detected by the sensors at critical steps, additional action needs to be taken in order to correct the subsequent errors. When each of these erroneous steps is encountered, similar actions as in *Case II* are to be taken through the respective mirror-nodes, without discarding the erroneous droplets. Finally, all such droplets produced at the end of the reaction paths are mixed together in a reservoir for mutual cancellation of all errors.

Example 3 Consider target-$CF = \frac{241}{512}$, for which the mix-split sequence is shown in Fig. 4.8 (*Case I*). For this target, Steps S_1, S_2, S_3, S_4, S_5 are known to be non-critical, and only Steps S_6, S_7, S_8 are critical for $\varepsilon = 7\%$. In the case of non-critical volumetric errors, no additional action will be taken by the controller as shown in Fig. 4.8. Now, if a volumetric split-error is detected online by a sensor at Step S_7, which is a critical step, two unequal volume droplets will be generated following this split operation. In order to cancel the error, the controller will execute Step S_8 and Step S_9 along with the original and the mirror-path and finally store all droplets in the common output reservoir to obtain the correct *CF* as shown in Fig. 4.8 (*Case II*).

Consider now the case of a double split-error (critical); let us assume that a split-error occurs at S_8 followed by an error at S_6. In order to cancel the generated error at S_6, the controller executes S_7' along the original path. Next, upon detection of a split-error at Step S_8, the controller, signaled by sensor feedback, will execute Step S_9 four times, in parallel, using the erroneous droplets along the mirror-paths, and mix the resulting four droplets together as shown in Fig. 4.8 (*Case III*). Thus, four erroneous droplets arriving from Steps S_9, S_9', S_9'', and S_9''' are combined to produce the desired error-free target-*CF*. Various error-recovery actions that are needed for roll-forward and rollback methods on the example shown in Fig. 4.8, are demonstrated in Section A.2 of the Appendix.

Recently, Bhattacharya *et al.* have proposed a highly-sensitive differential capacitive sensor that is capable of detecting very small amount of volumetric-error [16]. In this detection scheme, two component-droplets obtained after a split operation are sensed and fed to a differential comparator, which amplifies the difference-voltage, if any. If the volumetric-imbalance following a split operation exceeds certain desired threshold limit, the device outputs an error signal. When such a device is used, one can determine whether a volumetric-error occurring at a critical mix-split step, is indeed critical. If not, parallel roll-forward execution along mirror-paths need not be executed for error-cancellation at the target.

As for example, Fig. 4.9 shows the resulting *CF*-values obtained at the end of the reaction-path for target-$CF = \frac{241}{512}$ when various split-errors (%) are injected at

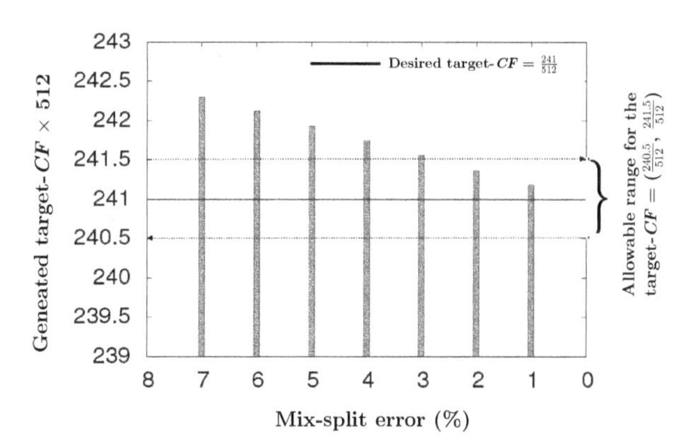

Figure 4.9: *CF*s obtained for various split-errors (ε) occurring at mix-split Step #6 (with permission from ACM [122])

mix-split Step #6, which is a critical step for 7% error. Note that for error $\leq 2\%$, this critical step becomes non-critical, since the resulting *CF* lies within the allowable error-range (Fig. 4.9). Hence, if the sensing scheme is capable of measuring the amount of volumetric-error online, correction measures need not be taken even for a critical step, when the error is beyond the threshold, based on which, the criticality was pre-computed.

4.4 DESIGNING AN LOC FOR IMPLEMENTING ECSP

In this section, we describe how an application-specific DMFB can be designed for implementing the above-mentioned error-correcting dilution algorithm (ECSP), and demonstrate the overall scheme with a simulator. Since the actuation sequence will depend on the feedback provided by on-chip sensors in real-time [48], the underlying task graph will consist of several control-flow sequences. In classical approaches to online recovery [4, 34, 48, 105], the flow graph needs to be re-executed from the latest checkpoint (rollback), when an error is signaled, and this action may involve additional re-synthesis overhead such as scheduling, placement, or droplet routing [62, 156, 179]. Thus, error recovery time may significantly increase. In the ECSP method, we obviate the need for re-synthesis/re-execution by adopting a roll-forward approach.

4.4.1 DESCRIPTION OF THE LAYOUT

We design an application-specific layout on 2D digital microfluidic chip for implementing the dilution scheme efficiently as shown in Fig. A.7 (Section A.4 of the Appendix). The reservoirs for dispensing sample and buffer fluids, and those for collecting target (output) and waste droplets are placed around the boundary of the chip. We use four (3×3) mixer modules, which may run in parallel when one or more critical errors are detected online. The sensor locations are shown in red color; differential capacitative sensors [16], or those based on a ring oscillator [59], or others [46] may be used to detect the presence of volumetric split-error. The cells shown in white are used for routing droplets. Note that routing of droplets must always satisfy fluidic constraints (static and dynamic) in space and time [23, 62].

4.4.2 SIMULATION OF ERROR-CORRECTING DILUTION

We load input dispensers with the given sample and buffer fluids, and assume that the dispense operations are free from any error, i.e., they will emit a unit-volume droplet (1X size) as needed. Dispense errors, if any, may be corrected before emission, as described in Section 4.2.1. Given the target-*CF* C_t, and accuracy level n, we approximate C_t as an n-bit binary fraction and determine the reaction path by running the *TWM* algorithm [163]. We then identify the mix-split steps that are critical for this

C_t by invoking Procedure 1. To avoid any unintentional droplet mixing, synchronization among operations is necessary such that no fluidic constraints are violated, e.g., if a droplet D occupies the cell at (x,y) at time t, then no other droplets can stay at any one of its eight neighbouring cells during $t-1$ to $t+1$ while routing. In the proposed layout, each droplet has a dedicated routing path between a source-destination pair. Similarly, all four mixer modules can run independently when needed, and each of them has a separate sensor module. When the required droplets enter to a mixer module, mix operation is performed for several clock cycles as specified, followed by a split operation. If the corresponding mix-split step is known to be non-critical, one of the two split droplets is used for the next operation as usual, and the other one is routed to the waste reservoir. Otherwise, if the step is critical, the two resulting droplets are routed to the nearest sensor node to check whether or not a volumetric split-error has indeed occurred. In no error is observed, the previous action as in the non-critical case is taken. Otherwise, the two erroneous droplets are transported, one each, to two different mixer modules for further processing of the remaining reaction paths in parallel (through the original and mirror nodes, respectively, as described in Procedure 2). Fig. A.8 depicts some snapshots that capture the process of detecting a volumetric split-error. Given the input specification, the simulation tool that can execute the complete procedure and compute assay-completion time *in-silico*.

We have built the simulator in Python language for executing a biochemical assay that is specified as input, using a high-level fluidic language (HLFL), which is similar to BioCoder [48]. The control-flow sequence of HLFL can capture sensor feedback and activate associated actuation sequences as needed. The tool simulates each clock cycle starting from the initial step of dispensing input fluids on the chip to the final step of producing target droplets of desired concentration factor, taking care of real-time occurrence of critical split-errors, if any. It also estimates the completion time of the given dilution assay on the basis of the time required by each operation, such as dispense, transport, mix, split, or detect [155]. These data may change from time to time depending on the nature of the assay or when the microfluidic platform/technology undergoes some shift or scaling [weblink].

4.5 EXPERIMENTAL RESULTS

In order to evaluate the performance and effectiveness of ECSP, we conduct simulation experiments on various synthetic examples as well as on some real-life test cases. Given target-*CF* C_t, we inject volumetric split-errors randomly at different mix-split steps on the reaction path as determined by *TWM*. For the purpose of comparison, we also consider a baseline scenario where we simulate the dilution procedure without assuming any corrective measure, and the resulting percentage error (%) in the target-*CF* is computed. The experiment is repeated on a number of *CF*s for $n = 9$.

We consider some real-life test cases where diluted fluids are needed as a preprocessing step. For example 70% ($\approx \frac{358}{512}$) ethanol is required as a sample in Glucose Tolerance Test in mice (for deciding how quickly exogenous glucose is cleared from blood), and in E.coli Genomic DNA Extraction (for purifying the genomic DNA without using commercial kits) [17]. Also, 95% ($\approx \frac{486}{512}$) ethanol is required for total RNA extraction from worms (with or without using commercial RNA extraction kits) and for Nuclear Staining of live worm [17]. A sample of 10% ($\approx \frac{51}{512}$) Fetal Bovine Serum (FBS) is required for *in-vitro* culture of human Peripheral Blood Mono-nuclear Cells (PBMCs) [17]. The simulation is carried out repeatedly for all these test cases for a number of volumetric split-errors inserted randomly at different mix-split steps.

Table 4.4
Average concentration error (%) for real-life test cases.

Target-*CF*	Single split-error				Double split-error				Approximation limit (%)
	Baseline		ECSP		Baseline		ECSP		
	Mean	Std	Mean	Std	Mean	Std	Mean	Std	
$\frac{358}{512}$	0.32	0.47	0.02	0.04	0.57	0.60	0.01	0.01	0.14
$\frac{51}{512}$	2.22	2.25	0.12	0.14	4.47	2.98	0.11	0.21	0.98
$\frac{486}{512}$	0.18	0.12	0.01	0.01	0.38	0.16	0.01	0.01	0.10

Table 4.5
Average reactant and mix-split overhead (%) for real-life test cases.

Target-*CF*	Single split-error						Double split-error					
	Sample		Buffer		Mix-split		Sample		Buffer		Mix-split	
	Mean	Std	Mean	Std	Mean	Std	Mean	Std	Mean	Std	Mean	Std
$\frac{358}{512}$	0.38	0.70	0.50	0.71	0.88	1.36	0.79	1.01	1.11	1.14	1.89	2.08
$\frac{51}{512}$	0.11	0.31	1.00	1.25	1.11	1.45	0.22	0.42	2.28	2.19	2.50	2.44
$\frac{486}{512}$	1.50	1.32	0.38	0.70	1.88	1.83	3.68	2.66	0.79	1.01	4.46	3.36

Tables 4.4 and 4.6 show the results on real-life and synthetic test cases, respectively, for both baseline (*TWM*) and ECSP scheme. A number of single and double volumetric split-errors are inserted on the various mix-split steps randomly; the mean and standard deviation of the error in target-*CF* are reported in these tables. It can be observed from Tables 4.4 and 4.6 that the error in target concentration exceeds the allowable accuracy (approximation) limit for the baseline approach. In other words,

such errors indeed affect the desired target-*CF*s. However, ECSP efficiently limits the error in target concentration within the desired accuracy level. We also report additional consumption of sample/buffer droplets and mix-split operational overhead needed for correcting single and double volumetric split-errors for real-life and synthetic test cases, in Table 4.5 and 4.7, respectively.

Table 4.6
Average concentration error (%) for synthetic test cases.

Target-*CF*	Single split-error				Double split-error				Approximation limit (%)
	Baseline		ECSP		Baseline		ECSP		
	Mean	Std	Mean	Std	Mean	Std	Mean	Std	
$\frac{13}{512}$	3.93	2.41	0.44	0.83	8.01	3.27	0.38	0.91	3.84
$\frac{75}{512}$	1.31	1.77	0.15	0.18	2.64	2.33	0.21	0.27	0.66
$\frac{391}{512}$	0.47	0.55	0.015	0.03	0.95	0.73	0.01	0.03	0.13
$\frac{451}{512}$	0.47	0.55	0.01	0.01	1.00	0.73	0.01	0.01	0.11

Table 4.7
Average reactant and mix-split overhead (%) for synthetic test cases.

Target-*CF*	Single split-error						Double split-error					
	Sample		Buffer		Mix-split		Sample		Buffer		Mix-split	
	Mean	Std	Mean	Std	Mean	Std	Mean	Std	Mean	Std	Mean	Std
$\frac{13}{512}$	0.0	0.0	1.56	1.89	1.56	1.89	0.0	0.0	3.56	3.33	3.56	3.33
$\frac{75}{512}$	0.11	0.31	0.78	1.31	0.89	1.59	0.22	0.42	1.67	1.99	1.89	2.34
$\frac{391}{512}$	0.78	0.92	0.33	0.67	1.11	1.45	1.81	1.70	0.69	0.97	2.50	2.44
$\frac{451}{512}$	1.33	1.33	0.33	0.67	1.67	1.83	3.19	2.59	0.69	0.97	3.89	3.27

We have also performed simulation experiments on the entire range of 511 target-*CF*s for accuracy level $n = 9$, and report the mixing and checking overhead (required by ECSP, compared to the baseline *TWM* algorithm). In these experiments, single

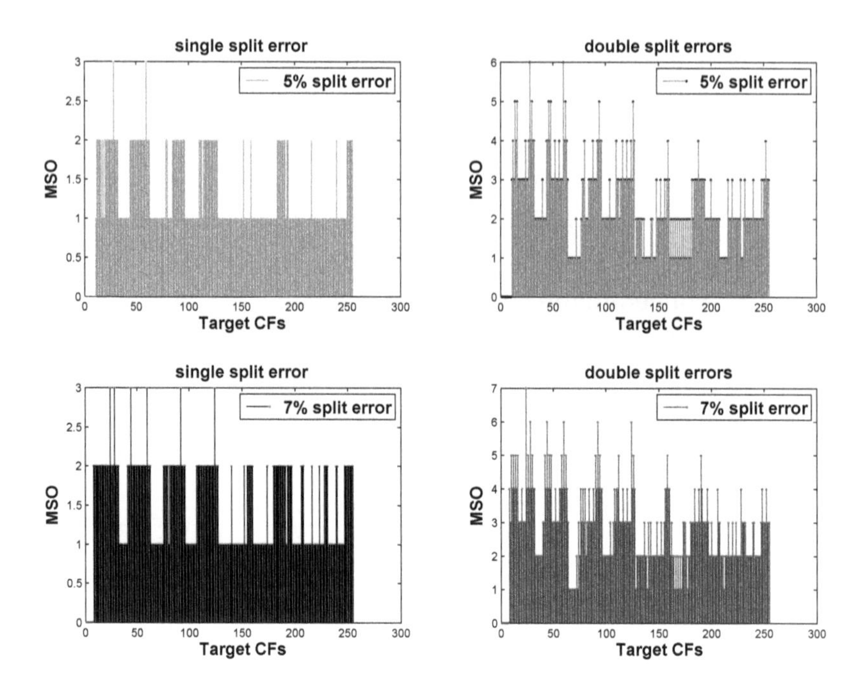

Figure 4.10: Mix-split overhead (MSO) for different values of volumetric errors (with permission from ACM [122])

and double volumetric split-errors have also been randomly inserted on all mix-split steps. Results are reported in Fig. 4.10 (for mixing overhead), and in Fig. 4.11 (for checking overhead) for various amount of split-error $\varepsilon = 5\%, 7\%$. Note that we have shown results up to $\frac{256}{512}$, because for the remaining range of *CF*s, the results will be similar (by Corollary 2). The average values of mixing and checking overhead (for ECSP, compared to the baseline) over the entire range of *CF*s are shown in Table 4.8.

Table 4.8
Average Mixing and checking overhead for accuracy level = 9.

Error (%)	Single split-error		Double split-error	
	Mix	Checkpoint	Mix	Checkpoint
3%	0.59	0.29	1.29	0.62
5%	0.88	0.50	1.99	1.08
7%	1.08	0.65	2.47	1.41

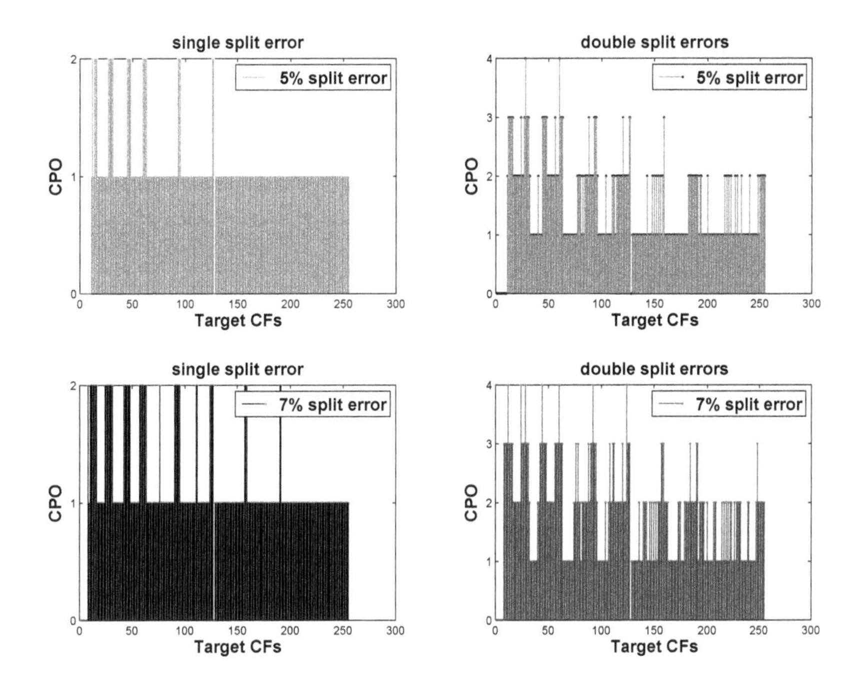

Figure 4.11: Checkpoint overhead (CPO) for different values of volumetric error (with permission from ACM [122])

We have also evaluated the performance of ECSP on some important assays that are widely used in clinical diagnostics, proteomics, and drug discovery, such as protein assay [91]. As described in the Bradford reaction [150], the protocol for a generic-droplet-based colorimetric protein assay is as follows. First, a sample droplet, such as serum or some other psychological fluid containing protein, is dispensed into the biochip. For diluting the sample, a buffer of 1M NaOH solution is used. On-chip dilution is performed using a multiple hierarchy of binary mixing/splitting steps, called the interpolating serial dilution method [40], which produces a concentration factor of $\frac{1}{2^k}$ after k steps of serial (exponential) dilution. If a volumetric split-error occurs at any of the mix-split operations, it will affect the subsequent operations of the protein assay, and as a result, the final concentration factor obtained at the end will be different from the desired one. As noted by Ren *et al.* [40], the error due to serial dilution may increase exponentially with the number of serial steps.

We consider the protein dilution assay as implemented on a DMFB [40, 150], and simulate the performance of ECSP, and an earlier rollback-based error-recovery scheme implemented with Biocoder [48], for comparison. For various randomly injected split-errors, we compute average completion time for both schemes, and the results are shown in Table 4.9, considering dispense time as 7s, dilution time 8s,

Table 4.9
Time needed by roll-forward and rollback approaches for error correction when n = 10.

Method (%)	Error-free baseline	one error	two errors	three errors	four errors
Roll-forward (ECSP) [122]	115.75	115.75	115.75	115.76	115.76
Rollback [48][1]	140.95	148.07	155.19	162.31	169.43

[1] without considering the criticality of errors (rollback approach)

detection time 5s, and routing time per electrode 10^{-2}s. Although the simulator in [48] may perform a fixed number of re-merge and re-split when an error is detected, for the purpose of comparison, we have considered the best case by assuming that the error will be corrected at the first re-merge/re-split attempt. Furthermore, ECSP allows parallel mix-split operations that may be necessitated when correction of multiple errors is required. Table 4.9 shows that ECSP can achieve error-tolerance in no extra time provided a sufficient number of mixer modules are available on-chip for parallel execution of all mirror-paths as discussed in Section 4.3.2.

4.6 CONCLUSIONS

In this chapter, we have described an error-tolerant scheme for sample preparation (ECSP) suitable for implementation with digital microfluidic lab-on-chips. This approach is resilient to multiple volumetric split-errors, the effects of which are mutually canceled at the target by adopting a novel roll-forward strategy. Consequently, a sample of correct concentration factor is produced as output as desired, regardless of the occurrence of any volumetric split-error on the reaction path. In contrast to rollback approaches, ECSP provides guaranteed and complete error-tolerance. We also present the layout design of a digital microfluidic biochip for efficient implementation of the scheme. Experimental results show that even in the presence of a single or double split-error, assay-completion time does not increase when this chip is used to execute the assay. In order to correct additional number of errors, one may have to pay some penalty, only for critical errors, in completion time or in space (i.e., more on-chip mixers are needed). Simulation results reveal that the overhead in terms of additional mix-split steps and sensor actuation increases with the amount of volumetric split-error. Also, in ECSP, a significant number of split-errors are classified as being non-critical, and therefore, unlike existing methods, no action is thus needed to correct them. Simulation results on several synthetic and real-life test cases demonstrate the efficacy and superiority of ECSP compared to rollback approaches.

Note that ECSP is based on *TWM*-type of algorithms [163], which produce one waste droplet at every mix-split step. The roll-forward strategy presented in this chapter utilizes these waste droplets when a critical volumetric split-error is sensed. The method can be possibly generalized for other similar algorithms such as *REMIA* [60]. However, the roll-forward method cannot be easily adopted for another class of dilution algorithms such as *DMRW* [131], or *IDMA* [132], where some of these waste droplets are reused in the dilution assay that relies on an iterated binary-search method. Development of new techniques for error-correction that will be suitable for such sample-preparation algorithms, may be investigated further.

5 Effect of Volumetric Split-Errors on Target-Concentration

In this chapter, an analysis of multiple volumetric split-errors and their effects on target-CFs are summarized [123]. Split-errors may unexpectedly occur in any mix-split step of the mixing-path during sample preparation, thus affecting the concentration factor (CF) of the target-droplet (see Section 4.2.3). Moreover, due to the unpredictable characteristics of fluidic-droplets, a daughter droplet of larger or smaller size may be used following an erroneous split operation[1] on the mixing-path. Although a number of cyber-physical-based approaches were proposed for error-recovery, they do not provide any guarantee on the number of rollback iterations that are needed to rectify the error. Thus, most of the prior approaches to error-recovery in biochips are non-deterministic in nature. The approach based on *ECSP* presented in the last chapter, on the other hand, performs error-correction in a deterministic sense; however, it assumes only the presence of single split-errors while classifying them as being critical or non-critical. *ECSP* does not consider the possibility of multiple split-errors during classification. Furthermore, in a cyber-physical settings, it requires some additional time for sensing the occurrence of a *critical* error, if any, at every such step. Hence, when the number of critical errors becomes large, sensing time may outweigh the gain obtained in roll-forwarding assay-time, and as a result, we may need a longer overall execution time.

5.1 ERROR-RECOVERY APPROACHES: PRIOR ART

Earlier approaches perform error-recovery operations by repeating the concerned portions of the bioassay [109] for producing the target concentration factor within the allowable error-range. For example, all mix-split operations and dispensing operations of the initial sequencing graph (shown in Fig. 5.1) were re-executed when an error is detected at the end. However, the repetition of such experiments leads to wastage of precious reagents and hard-to-obtain samples, and results in longer assay-completion time.

[1]depending on the selection of the erroneous droplet (larger or smaller volume) to be used in a subsequent step.

DOI: 10.1201/9781003219651-5

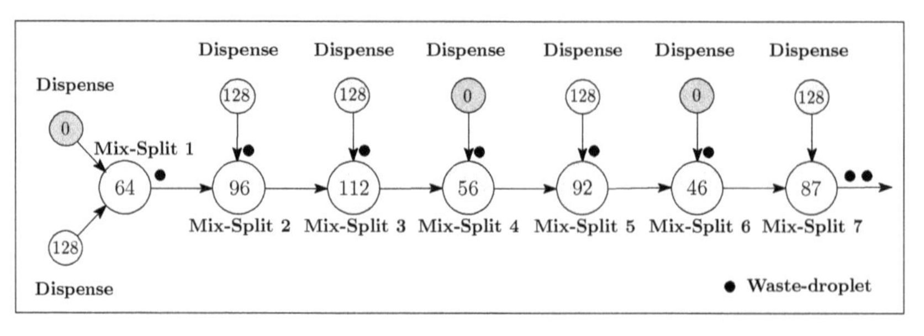

Figure 5.1: Initial sequencing graph for generating target-$CF = \frac{87}{128}$

5.2 CYBER-PHYSICAL TECHNIQUE FOR ERROR-RECOVERY

In order to avoid repetitive execution of biochemical experiments, cyber-physical DMFBs were proposed for obtaining the desired outcome in the presence of errors [105]. The recovery scheme is shown in Fig. 5.2.

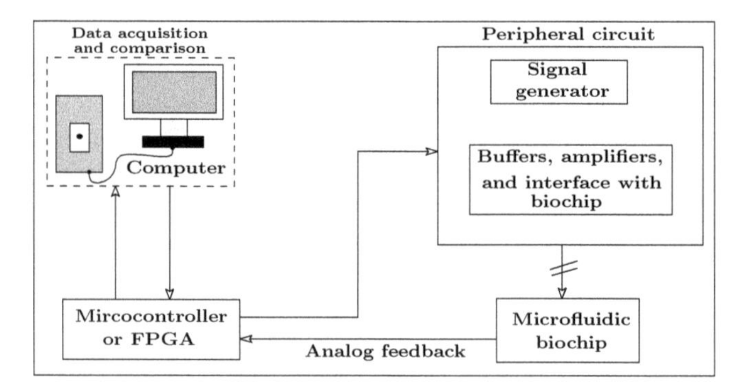

Figure 5.2: Schematic of a cyber-physical error-recovery system

It consists of the following components: a computer, a microcontroller or an FPGA, a peripheral circuit, and the concerned biochip. Two interfaces are required for establishing the connection between control software and hardware of the microfluidic system. The first one is required for converting the output signal of the sensor to an input signal that feeds the control software installed on the computer. The second interface transforms the output of the control software into a sequence of voltage-actuation maps that activate the electrodes of the biochip. The error-recovery operation is executed by the control software running in the back-end.

5.2.1 COMPILATION FOR ERROR-RECOVERY

In cyber-physical-based DMFBs, one has to monitor the output of the intermediate mix-split operations at designated checkpoints using on-chip sensors. The original actuation sequences are interrupted when an error is detected during the execution of a bioassay. At the same time, the recovery actions, e.g., the re-execution of corresponding dispensing and mixing operations is initiated to remedy the error. However, the error-recovery operations will have to be translated into electrode-actuation sequences in real-time. The compilation of error-recovery actions can either be performed before the actual execution of the bio-assay or during its execution. So, depending on the compilation-time of operations, error-recovery approaches can be divided into two categories: i) off-line (at design time), and ii) online (at run time).

In the off-line approach, the errors that might occur (under the assumed model) during the execution of a bio-assay are identified, and compilation is performed to pre-compute and store the corresponding error-recovery actuation sequences which will provide an alternative schedule. However, this approach can be used to rectify only a limited number of errors (≤ 2) since a very large-size controller memory will be required to store the recovery sequences for possible errors [106]. On the other hand, in the online approach, appropriate actions are carried out depending on the feedback given by the sensor. Compilation of error-recovery actions into electrode-actuation sequences is performed only at run-time.

5.2.2 WORKING PRINCIPLE OF CYBER-PHYSICAL-BASED DMFBS

For both off-line and online approaches stated above, cyber-physical DMFBs perform error-recovery operations as follows. During actual execution of the bio-assay, a biochip receives control signals from the software running on the computer system. At the same time, the sensing system of the biochip sends a feedback signal to the software by processing it using field-programmable gate array (FPGA), or application-specific integrated circuit (ASIC) chips. If an error is detected by a sensor, the control software immediately discards the erroneous-droplets for preventing error-propagation, and performs necessary error-recovery operations (i.e., the corresponding actuation sequences are determined online/off-line).

In order to produce the correct output, the outcomes of intermediate mix-split operations are verified using on-chip sensors suitably placed at designated checkpoints. For example, in Fig. 5.3, the outcomes of *Mix-split* 4 and *Mix-split* 7 are checked by a sensor. When an error is detected, a portion of the bio-assay is re-executed. For instance, the operations shown within the blue box in Fig. 5.3 are re-executed when an error is detected at the last checkpoint. Note that the accuracy of a cyber-physical system also depends on the sensitivity of sensors. Unfortunately, due to cost constraints, only a limited number of sensors can be integrated on a DMFB [105]. Additionally, in order to check the status of intermediate droplets, they need to be routed to a designated sensor location on the chip. This may introduce significant latency to the overall assay-completion time (Fig. 5.4). As a result, most of the cyber-physical-based error-recovery methods for sample preparation turn out to be expensive in terms of assay-completion time and reagent cost.

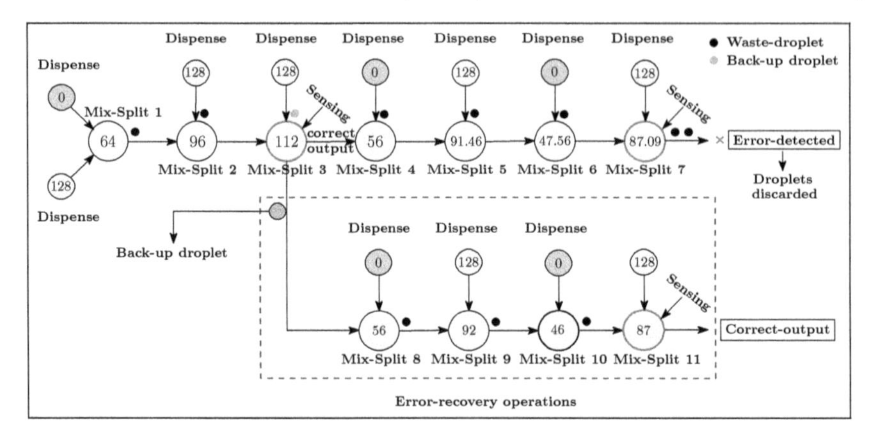

Figure 5.3: Sample preparation using cyber-physical error-recovery LoC system

To summarize, cyber-physical error-recovery methods suffer from the following shortcomings:

- They are expensive in terms of assay-completion time and reagent-cost. Hence, they are unsuitable for field deployment and point-of-care testing in resource-constrained areas.
- They may fail to provide any guarantee on the number of rollback attempts, i.e., how many iterations will be required to correct the error. Hence, error-recovery becomes non-deterministic.
- Each additional component used in the design may become a possible source of failure, which ultimately reduces the reliability of the biochip.

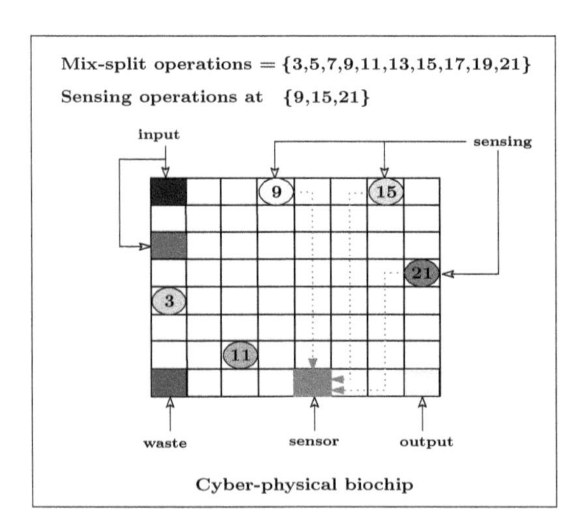

Figure 5.4: Routing of droplets for sensing operation in a cyber-physical biochip

5.3 EFFECT OF SPLIT-ERRORS ON TARGET-*CF*s

Generally, in the (1:1) mixing model, two 1X-volume daughter-droplets are produced after each mix-split operation. One of them is used in the subsequent mix-split operation and another one is discarded as waste droplet or stored for later use [163] (see Fig. 5.1). An erroneous mix-split operation may produce two unequal-volume droplets. Unless an elaborate sensing mechanism is used, it is not possible to predict which one of the resulting droplets (smaller/larger) is being used in the subsequent mix-split operation. Moreover, their effect on the target-*CF* becomes more complex when multiple volumetric split-errors occur in the mix-split path.

5.3.1 SINGLE VOLUMETRIC SPLIT-ERROR

In order to analyze the effect of single volumetric split-error on the target-*CF*, we perform experiments with different erroneous droplets and present relevant results in this section. We assume an example target-$CF = \frac{87}{128}$ of accuracy level = 7. The mix-split sequence that needs to be performed using *twoWayMix* algorithm [163] for generating the target-*CF* is shown in Fig. 5.1.

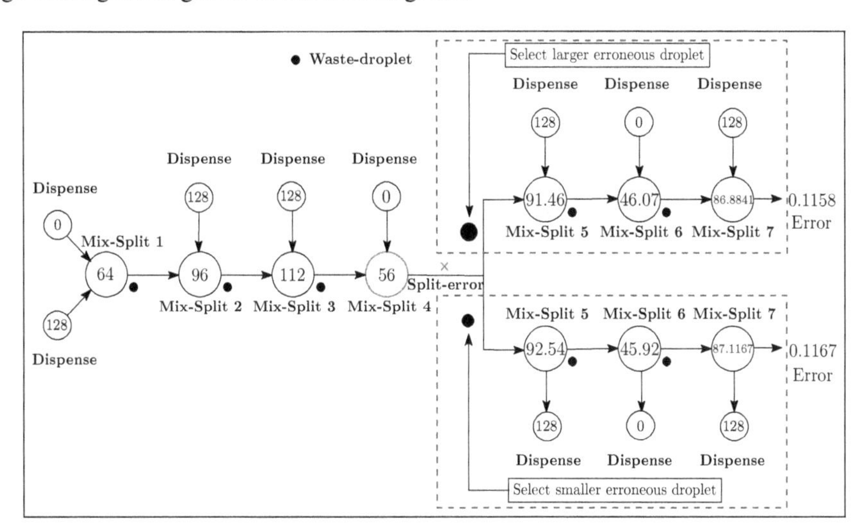

Figure 5.5: Effect of choosing larger-/smaller-volume erroneous droplet on target-$CF = \frac{87}{128}$

Let us consider the scenario of injecting volumetric split-error at mix-split Step 4. Two unequal-volume daughter droplets are produced after this step when a split-error occurs. The effect of the erroneous droplet on the target-*CF* depends on the choice of the daughter-droplet to be used next. For example, the effect of 3% volumetric split-error (at Step 4) on the target-$CF = \frac{87}{128}$ is shown in Fig. 5.5. The effect of two errors on the target-*CF* (when the larger or smaller volume droplet is used) is also shown in Fig. 5.5. The blue (green) box represents the scenario when the next operation is executed with the larger (smaller) erroneous droplet. It has been seen from Fig. 5.5 that the *CF*-error in the target increases when the smaller erroneous droplet is used in the mixing path compared to the use of the larger one. We perform further experiments for finding the effect of erroneous droplets on a target-*CF*. We report the results for volumetric split-error 3%, 5% and 7% occurring on the mixing path, in Table 5.1. We observe that the *CF*-error in the target-droplet exceeds the error-tolerance limit in all cases when a volumetric split-error occurs in the last but one step. Moreover, the *CF*-error in the target-*CF* increases when the magnitude of volumetric split-error increases.

5.3.2 MULTIPLE VOLUMETRIC SPLIT-ERRORS

To this end, we have analyzed the effect of single volumetric split-error on a target-*CF*. However, due to unpredictable characteristics of fluid droplets, such split-errors may occur in multiple mix-split steps of the mixing path, and as a result, they may change the *CF* of the desired target-droplet significantly. Moreover, volumetric split-errors may occur with any combination of signs (use of larger or smaller droplet following a split step) on the mixing path during sample preparation. We derive expressions that capture the overall effect of such errors on the target-*CF*.

Let ε_i indicate the percentage of the volumetric split-error occurring at the ith mix-split step. A natural question in this context is: "How is the *CF* of a target-droplet affected by multiple volumetric split-errors $\{\varepsilon_1, \varepsilon_2, \ldots, \varepsilon_{i-1}\}$ occurring on different mix-split steps in the mixing path during sample preparation?"

Let us consider the dilution problem for generating target-$CF = C_t$ using *twoWayMix* [163] as shown in Fig. 5.6. Here, O_i represents the ith (1:1) mix-split step, C_i is the resulting *CF* after the ith mix-split step, and r_i is the *CF* of the source (100% for sample, 0% for buffer) used in i^{th} mix-split operation. Without loss of generality, let us assume that a volumetric split-error ε_i has occurred after the ith mix-split step of the mixing path, i.e., two daughter-droplets of volume $1+\varepsilon$ and $1-\varepsilon$, $\varepsilon > 0$, have been produced following a split. Initially, sample and buffer are mixed at the first mix-split step (O_1). After this mixing operation, the *CF* and volume of the resulting droplet become: $C_1 = \frac{P_0 \times (1 \pm \varepsilon_0) + 2^{-1} \times r_0}{Q_0 \times (1 \pm \varepsilon_0) + 2^{-1}}$ and $V_1 = \frac{Q_0 \times (1 \pm \varepsilon_0) + 2^{-1}}{2^0}$, respectively, where $P_0 = Q_0 = \frac{1}{2}$, $\varepsilon_0 = r_0 = 0$. Note that $r_i = 1$ (0) indicates whether a sample (buffer) is used in the ith mix-split step of the mixing path. Furthermore, the sign $+$ ($-$) in the expression indicates whether a larger (smaller) droplet is used in the next mix-split step followed by a split operation.

A volumetric split-error may occur in one or more mix-split operations of the mixing path while preparing the target-*CF*. For example, volumetric split-errors $\{\varepsilon_1, \varepsilon_2, \ldots, \varepsilon_6\}$ may occur, one after another, in mix-split operations $\{O_1, O_2, \ldots, O_6\}$ as shown in Fig. 5.6. In Table 5.2, we report the volume and concentration of the resulting daughter-droplets after each mix-split operation when all preceding steps suffer from split-errors.

Hence, for the occurrence of multiple volumetric split-errors, say $\{\varepsilon_1, \varepsilon_2, \varepsilon_3, \ldots, \varepsilon_{i-2}, \varepsilon_{i-1}\}$ at mix-split steps $\{O_1, O_2,, O_3, \ldots, O_{i-2}, O_{i-1}\}$, the *CF* and volume of the generated target-droplet after the final mix-split operation can be computed using

Table 5.1
Impact on target-$CF = \frac{87}{128}$ for different volumetric split-errors.

Volumetric split-error = 3%.				
Erroneous mix-split step	Selected-droplet		Target-$CF \times 128$*	Within error-tolerance limit?
	Larger	Smaller		(CF-error $\times 128 < 0.5$?)
1	✓	×	86.98	Yes
1	×	✓	87.01	Yes
2	✓	×	87.01	Yes
2	×	✓	86.99	Yes
3	✓	×	87.04	Yes
3	×	✓	86.95	Yes
4	✓	×	86.88	Yes
4	×	✓	87.12	Yes
5	✓	×	87.04	Yes
5	×	✓	86.96	Yes
6	✓	×	86.39	No
6	×	✓	87.62	No

Volumetric split-error = 5%.				
Erroneous mix-split step	Selected-droplet		Target-$CF \times 128$	Within error-tolerance limit?
	Larger	Smaller		(CF-error $\times 128 < 0.5$?)
1	✓	×	86.98	Yes
1	×	✓	87.02	Yes
2	✓	×	87.01	Yes
2	×	✓	86.98	Yes
3	✓	×	87.08	Yes
3	×	✓	86.92	Yes
4	✓	×	86.81	Yes
4	×	✓	87.19	Yes
5	✓	×	87.06	Yes
5	×	✓	86.94	Yes
6	✓	×	86.00	No
6	×	✓	88.05	No

Volumetric split-error = 7%.				
Erroneous mix-split step	Selected-droplet		Target-$CF \times 128$	Within error-tolerance limit?
	Larger	Smaller		(CF-error $\times 128 < 0.5$?)
1	✓	×	86.97	Yes
1	×	✓	87.03	Yes
2	✓	×	87.02	Yes
2	×	✓	86.98	Yes
3	✓	×	87.11	Yes
3	×	✓	86.89	Yes
4	✓	×	86.73	Yes
4	×	✓	87.27	Yes
5	✓	×	87.09	Yes
5	×	✓	86.91	Yes
6	✓	×	85.61	No
6	×	✓	88.49	No

*: Results are rounded up to two decimal places.

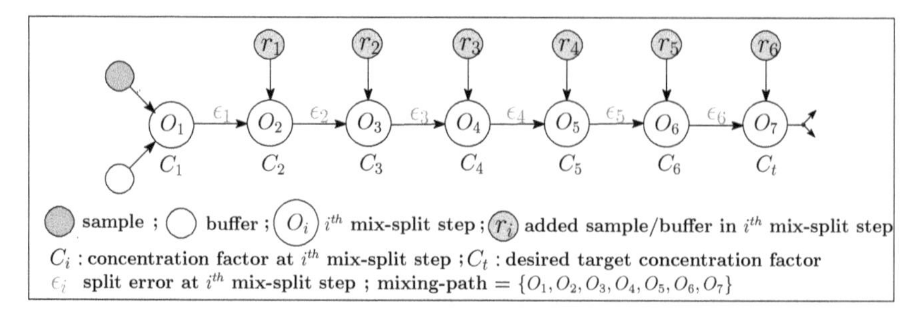

Figure 5.6: Mix-split operations for generating target-CF = C_t with accuracy n = 7 (with permission from IEEE [124])

the following expressions:

$$C_i = \frac{P_{i-1} \times (1 \pm \varepsilon_{i-1}) + 2^{i-2} \times r_{i-1}}{Q_{i-1} \times (1 \pm \varepsilon_{i-1}) + 2^{i-2}} \tag{5.1}$$

$$V_i = \frac{Q_{i-1} \times (1 \pm \varepsilon_{i-1}) + 2^{i-2}}{2^{i-1}} \tag{5.2}$$

where $P_i = P_{i-1} \times (1 \pm \varepsilon_{i-1}) + 2^{i-2} \times r_{i-1}$ and $Q_i = Q_{i-1} \times (1 \pm \varepsilon_{i-1}) + 2^{i-2}$. In this way, the impact of multiple volumetric split-errors occurring on different mix-split steps of the mixing path on the target-CF can be pre-computed.

In order to find the effect of multiple volumetric split-errors on a target-CF, we perform several experiments. We continue with the example target-CF = $\frac{87}{128}$ with accuracy level = 7, and inject 7% volumetric split-error simultaneously at different mix-split steps of the mixing path. The effects of such split-errors are shown in Fig. 5.7. During simulation, we assume that the larger erroneous droplet is always used later when a split-error occurs in the mix-split path (i.e., ε is positive). For example, the effect of multiple 7% volumetric split-errors in Mix-Split Step 1 and Step 3 is shown in Fig. 5.7(b). Only the effect of three concurrent volumetric split-errors is shown in Fig. 5.7 (c). It has been observed that CF-error in the target-droplet rapidly grows to $\frac{0.08}{128}$ and $\frac{0.17}{128}$ when two or three such split-errors are injected in the mix-split path.

5.4 WORST-CASE ERROR IN TARGET-CF

So far we have analyzed the effect of multiple volumetric split-errors on a target-CF when a larger erroneous droplet is selected following each mix-split step. In reality, multiple volumetric split-errors may consist of an arbitrary combination of large and small daughter-droplets. Hence, further analysis is required to reveal the role of such random occurrence of volumetric split-errors and their effects on the target-CF.

In order to facilitate the analysis, we define "error-vector" as follows: An error-vector of length k denotes the sequence of larger or smaller erroneous droplets, which

Table 5.2
Impact of split-errors on resulting daughter-droplets.

Erroneous mix-split step (O_i)	Split-error	*CF*	*V*	Parameter values
$\{O_1\}$	$\{\varepsilon_1\}$	$C_2 = \frac{P_1\times(1\pm\varepsilon_1)+r_1}{Q_1\times(1\pm\varepsilon_1)+2^0}$	$V_2 = \frac{Q_1\times(1\pm\varepsilon_1)+2^0}{2}$	$P_1 = P_0\times(1\pm\varepsilon_0)+2^{-1}\times r_0$, $Q_1 = Q_0\times(1\pm\varepsilon_0)+2^{-1}$
$\{O_1, O_2\}$	$\{\varepsilon_1, \varepsilon_2\}$	$C_3 = \frac{P_2\times(1\pm\varepsilon_2)+2\times r_2}{Q_2\times(1\pm\varepsilon_2)+2}$	$V_3 = \frac{Q_2\times(1\pm\varepsilon_2)+2}{2^2}$	$P_2 = P_1\times(1\pm\varepsilon_1)+r_1$, $Q_2 = Q_1\times(1\pm\varepsilon_1)+2^0$
$\{O_1, O_2, O_3\}$	$\{\varepsilon_1, \varepsilon_2, \varepsilon_3\}$	$C_4 = \frac{P_3\times(1\pm\varepsilon_3)+2^2\times r_3}{Q_3\times(1\pm\varepsilon_3)+2^2}$	$V_4 = \frac{Q_3\times(1\pm\varepsilon_3)+2^2}{2^3}$	$P_3 = P_2\times(1\pm\varepsilon_2)+2\times r_2$, $Q_3 = Q_2\times(1\pm\varepsilon_2)+2$
...	...	...	...	...
...	...	...	...	...
$\{O_1,O_2,O_3,O_4,O_5,O_6\}$	$\{\varepsilon_1,\varepsilon_2,\varepsilon_3,\varepsilon_4,\varepsilon_5,\varepsilon_6\}$	$C_7 = \frac{P_6\times(1\pm\varepsilon_6)+2^5\times r_6}{Q_6\times(1\pm\varepsilon_6)+2^5}$	$V_7 = \frac{Q_6\times(1\pm\varepsilon_6)+2^5}{2^6}$	$P_6 = P_5\times(1\pm\varepsilon_5)+2^4\times r_5$, $Q_6 = Q_5\times(1\pm\varepsilon_5)+2^4$

are chosen corresponding to k mix-split errors in the mixing path. For example, an error-vector [+,ϕ,−,ϕ,ϕ,+] denotes volumetric split-error in Mix-Split Step 1, Step 3, and Step 6, where ϕ denotes no-error. In Step 1, the larger droplet is passed to the next step, whereas in Step 3, the smaller one is used in the next step, and so on. For k volumetric split-errors, 3^k error-vectors are possible. While executing actual mix-split operations, the target-*CF* can be affected by any one of them.

We perform simulation experiments for finding the effect of different error-vectors for target-*CF* = $\frac{87}{128}$. Initially, we observe the effect of three errors corresponding to the mix-split operations {Mix-Split 1, Mix-Split 3, Mix-Split 6} on the target-*CF* (for 7% split-error). See Fig. 5.8 for an example. We observe that the *CF*-error in the target-droplet increases noticeably for error-vectors [+,ϕ,+,ϕ,ϕ,+], [+,ϕ,−,ϕ,ϕ,+] and [−,ϕ,−,ϕ,ϕ,+] as depicted in Fig. 5.8 (a)–(c). It can be seen from Fig. 5.8 that the *CF*-error exceeds the error-tolerance limit (= $\frac{0.5}{128}$) in each case. Thus, the target-*CF* is affected badly for these error-vectors. We perform similar experiments with volumetric split-error 3% and found that *CF*-error decreases for all cases. Moreover, we perform simulation for revealing the effect of remaining error-vectors on the target-*CF* and report the errors for all possible error-vectors (# error-vectors = 8) in Table. 5.3. It has been observed that the *CF*-error exceeds allowable error-tolerance limit in all these cases. On the other hand, the maximum *CF*-error in the target-*CF* occurs for the error-vector [−,ϕ,+,ϕ,ϕ,−] which is $\frac{1.61}{128}$ (> error-tolerance limit). Note that volumetric split-error may also occur in the remaining mix-split

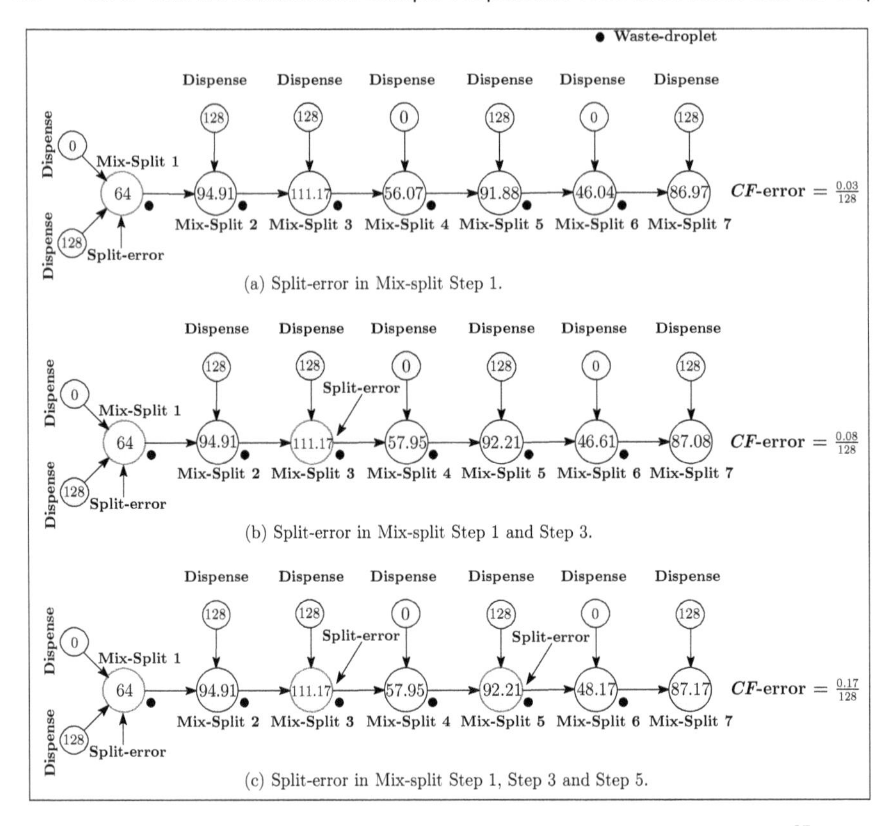

Figure 5.7: Effect of multiple volumetric split-errors on target-$CF = \frac{87}{128}$

steps, and they might impact the correctness of the target-*CF*.

We further perform experiments to find the effect of such volumetric split-errors on target-$CF = \frac{87}{128}$. The mix-split graph of target-$CF = \frac{87}{128}$ consists of 7 mix-split operations (see Fig. 5.1). During simulation, we inject split-error in each mix-split step of the mixing path except the final mix-split operation (only the volumes of two resulting target-droplets may change because of an error in the last mix-split step). Thus, there will be 64 possible error-vectors. We set split-error = +0.07 or -0.07, in each mix-split step, depending on the sign of the error in the corresponding position of vector. We report the results for some representative error-vectors for this target in Table 5.4. We see that the *CF*-error exceeds the allowable error-range in every case.

We also show the *CF*-error for all possible error-vectors in Fig. 5.9 for target-*CF* = $\frac{41}{128}$ and $\frac{87}{128}$ (complement of $\frac{41}{128}$) for demonstration purpose. We plot error-vectors (setting $+ \rightarrow 0$, $- \rightarrow 1$) along the X-axis, and arrange them from left-to-right following a gray-code, so that any two adjacent vectors are only unit Hamming distance apart. The Y-axis shows the corresponding values of *CF*-error$\times$128. Based on exhaustive simulation, we observe that the *CF*-error for both targets becomes maximum ($\frac{1.977}{128}$) for the error-vector $[-,+,+,-,+,-]$ (shown at the 57th position on the

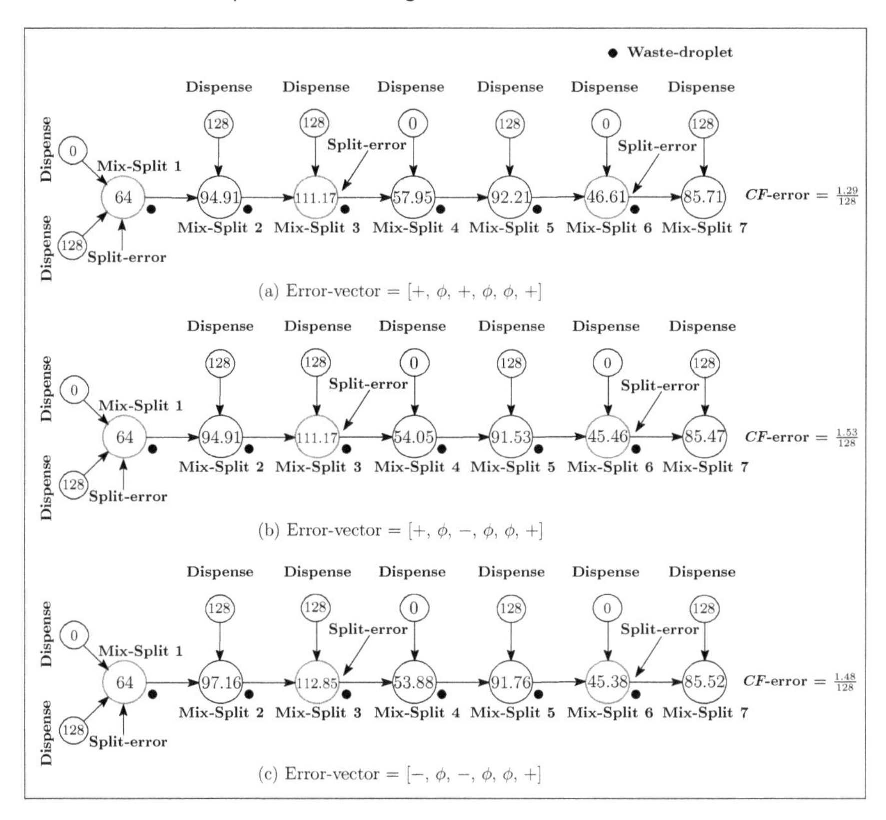

Figure 5.8: Effect of multiple volumetric split-errors on target-$CF = \frac{87}{128}$

X-axis). We notice that none of these outcomes lies within the safe-zone (i.e., error-tolerance limit). We also perform similar experiment for both target-*CF*s when split-error becomes 3% and observe that for 12 cases, the errors lies within the tolerance zone, and the maximum *CF*-error reduces to $\frac{0.84}{128}$ corresponding to the same error-vector [−,+,+,−,+,−] (Fig. 5.10). However, for target-$CF = \frac{17}{128}$, a large number of *CF*-errors = 29 (32) generated by all possible error-vectors of length 6 lie within the error-tolerance zone for 7% (3%) split-errors (see Fig. 5.11). Note that the magnitude of *CF*-errors decreases in each case when split-error reduces to 3%. We further perform simulation for measuring maximum *CF*-error generated for all target-*CF*s with accuracy level 7 (with 7% split-error). We plot the results in Fig. 5.12. We observe that the *CF*-error for target-$CF = \frac{63}{128}$ and $\frac{65}{128}$ becomes maximum ($\frac{4.12}{128}$) compared to those produced by other error-vectors. The error-vector [−, −, −, −, −, −] generates the maximum *CF*-error for both these target-*CF*s.

Table 5.3
Effect of multiple split-errors on target-*CF* = $\frac{87}{128}$ for 7% split-error.

Error-vector	Produced $CF \times 128$*	Produced CF-error $\times 128$	CF-error $\times 128 < 0.5$?
$[+, \phi, +, \phi, \phi, +]$	85.71	1.29	No
$[+, \phi, +, \phi, \phi, -]$	88.56	1.56	No
$[+, \phi, -, \phi, \phi, +]$	85.47	1.53	No
$[+, \phi, -, \phi, \phi, -]$	88.36	1.36	No
$[-, \phi, +, \phi, \phi, +]$	85.76	1.24	No
$[-, \phi, +, \phi, \phi, -]$	88.61	1.61	No
$[-, \phi, -, \phi, \phi, +]$	85.52	1.48	No
$[-, \phi, -, \phi, \phi, -]$	88.41	1.41	No

* Results are rounded up to two decimal places.

Table 5.4
Effect of multiple split-errors on the target-*CF* = $\frac{87}{128}$ for 7% split-error.

Error-vector	Produced $CF \times 128$*	Produced CF-error $\times 128$	CF-error $\times 128 < 0.5$?
$[+,+,+,+,+,+]$	85.58	1.42	No
$[+,-,+,+,+,+]$	85.53	1.47	No
$[+,-,-,+,+,+]$	85.26	1.74	No
$[+,-,+,+,-,+]$	85.08	1.92	No
$[-,+,-,-,+,-]$	88.78	1.78	No
$[-,+,+,-,-,-]$	88.82	1.82	No
$[-,+,-,-,-,-]$	88.64	1.64	No
$[-,-,-,-,-,-]$	88.61	1.61	No

* Results are rounded up to two decimal places.

5.5 MAXIMUM CF-ERROR: A JUSTIFICATION

Motivated by the need for a formal proof for generating maximum *CF*-error in the target-*CF*, we have performed rigorous theoretical analysis and further experiments to study the properties of *CF*-error in a target-*CF*. The following analysis, as shown below, reveals how the problem of error-tolerance can be handled in a more concrete fashion.

Consider a particular target-$CF = C_t$ and its dilution tree. Let the current mix-split step be i (other than the last step, where the occurrence of split-error does not matter),

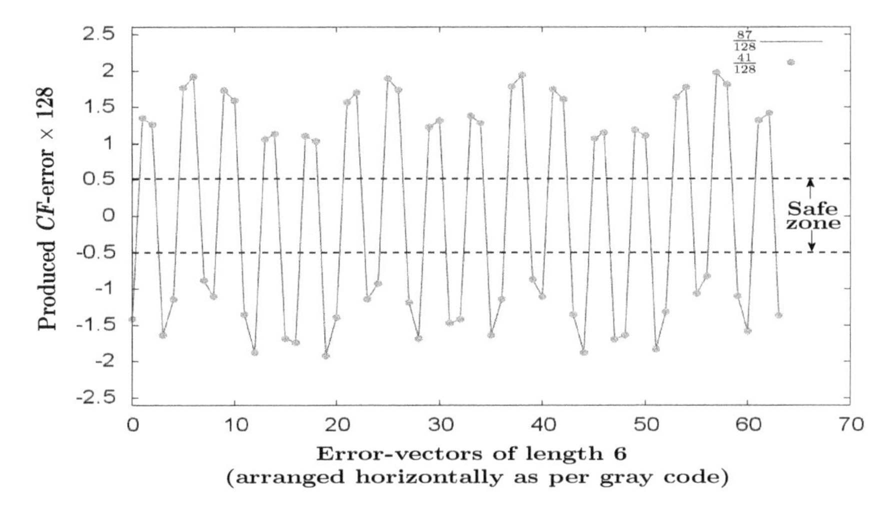

Figure 5.9: Value of (CF-error$\times$128) for all error-vectors with 7% split-error for target-CFs = $\frac{41}{128}$ and $\frac{87}{128}$

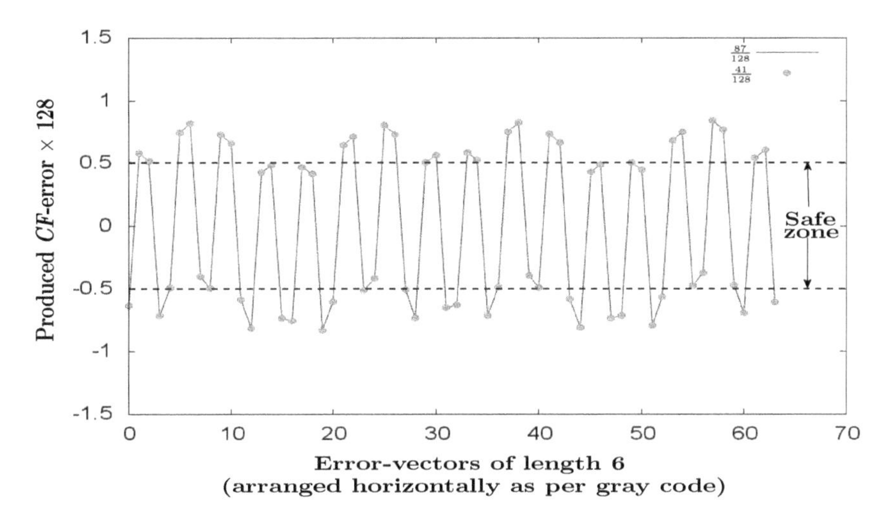

Figure 5.10: Value of (CF-error$\times$128) for all error-vectors with 3% split-error for target-CFs = $\frac{41}{128}$ and $\frac{87}{128}$

and the intermediate-CF arriving at the ith step be C_i. If a 1X sample (buffer) droplet is added in this step, it produces $CF = \frac{C_i+1}{2}$ (resp., $\frac{C_i}{2}$), assuming that the volume of the droplet arriving at the ith step is correct (1X).

Consider the first case, and assume that the droplet arriving at i suffers a volumetric split-error of magnitude ε at the previous step. Hence, after mixing with a sample droplet, the intermediate-CF will become: $\frac{C_i(1+\varepsilon)+1}{2+\varepsilon}$; the sign of ε is set to as

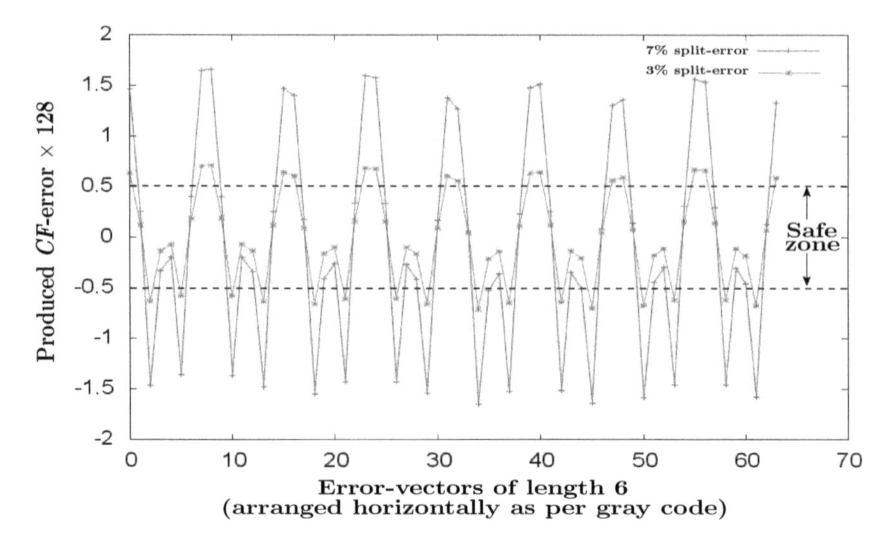

Figure 5.11: Value of (CF-error$\times 128$) for all error-vectors for 7% and 3% split-error for target-$CF = \frac{17}{128}$

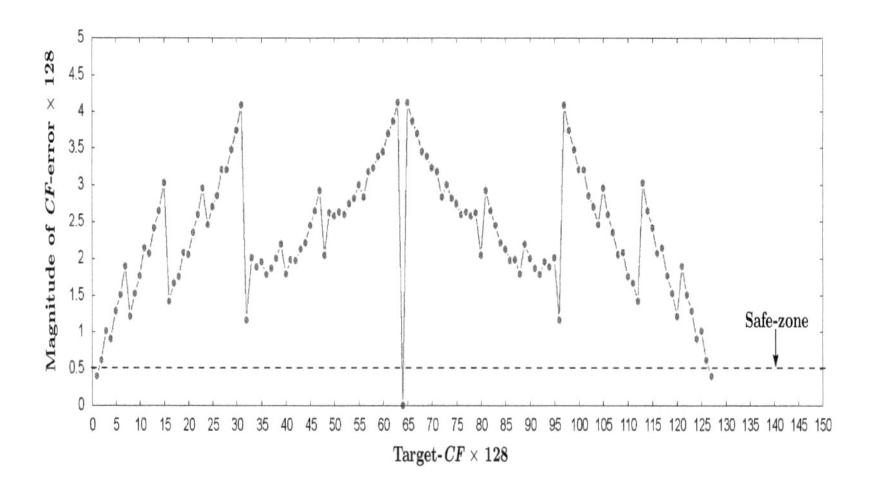

Figure 5.12: Maximum value of (CF-error$\times 128$) for all target-CFs with accuracy 7

positive (negative) when the incoming intermediate-droplet is larger (smaller) than the ideal volume 1X. Thus, the error (E_r) in the intermediate-CF becomes:

$$E_r = \frac{C_i + 1}{2} - \frac{C_i(1+\varepsilon)+1}{2+\varepsilon} = \frac{\varepsilon(1-C_i)}{4+2\varepsilon} \tag{5.3}$$

From Expression 5.3, it can be observed that the magnitude of E_r becomes larger when ε is negative, because a negative error reduces the value of the denominator.

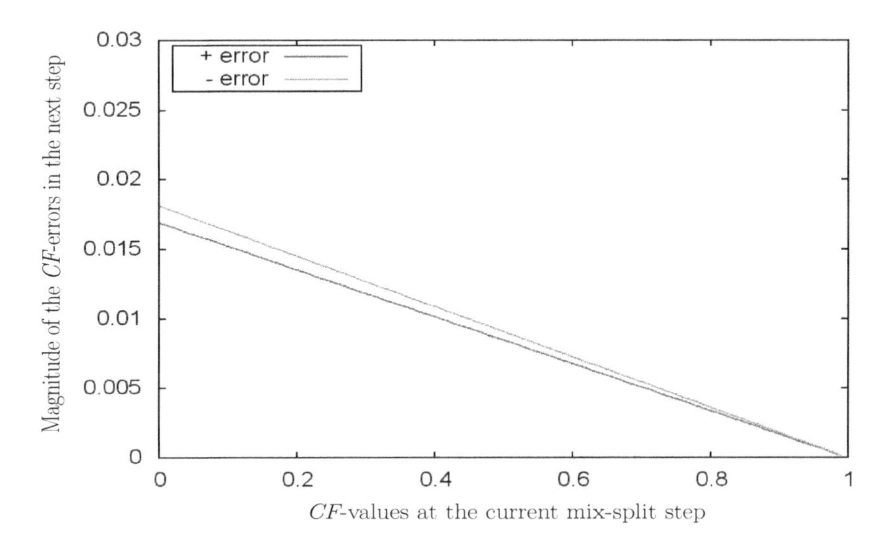

Figure 5.13: *CF*-error at the next mix-split step (for positive and negative single split-error)

In other words, the error in *CF* will become slightly more pronounced if a droplet of smaller-volume arrives at Step i compared to the case when a larger-volume droplet arrives at the mixer. In other words, the *effect of a split-error is not symmetrical*; however, since the volumes of the two daughters will be proportionately different as well, when they are mixed at the target, the error is canceled. We perform an experiment assuming volumetric error (7%), i.e., by setting ε = +0.07 or -0.07 in one mix-split step, for all values of intermediate-*CF*s. The corresponding results are shown in Fig. 5.13. It can be observed that a negative split-error always produces larger *CF*-error in the target-*CF* for a single split-error (error-vector of length 1). Similar effects will be observed when a buffer droplet is mixed at Step i.

We also perform simulation by varying C_i from 0 to 1, and ε from -0.07 to 0.07 in Expression 5.3 and calculate *CF*-errors. We report the results as 3-dimensional (3D) plots (with different views) in Figs. 5.14 and 5.15. We observe that simulation results favorably match with theoretical results (see Fig. 5.13). However, the impact on a target-*CF* becomes much more complicated when multiple split-errors are considered.

In order to demonstrate the intricacies, we have performed a representative analysis considering three consecutive split-errors. For simplicity, let us assume that an error of magnitude ε is injected in each of these three mix-split steps. Generalizing Expression 5.3, we can show that the corresponding *CF*-error observed after three steps will be:

$$E_r = \frac{((((C_i(1+\varepsilon)+r_1)(1+\varepsilon)+r_2))(1+\varepsilon)+r_3)}{(((2+(2+\varepsilon)(1+\varepsilon)))(1+\varepsilon)+4)} - \frac{(C_i+r_1+r_2+r_3)}{8} \quad (5.4)$$

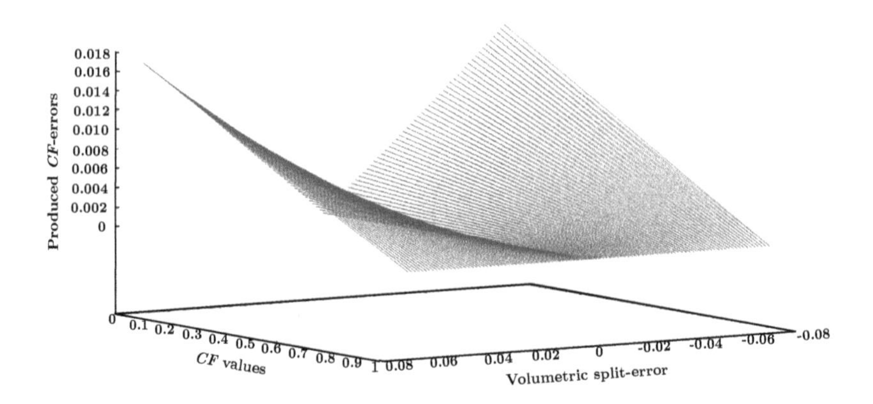

Figure 5.14: *CF*-error at the next mix-split step (for positive and negative single split-error)

where r_i = 1 (for sample droplet)
= 0 (for buffer droplet), for Step i, i = 1, 2, 3.

As before, we assume that ε = +0.07 or -0.07, and since we have three consecutive split steps, we have eight possible combinations of such error-vectors $[\phi^{\alpha}, -, -, -, \phi^{\beta}]$, $[\phi^{\alpha}, -, -, +, \phi^{\beta}]$, ... , $[\phi^{\alpha}, +, +, +, \phi^{\beta}]$ for a given combination of r_1, r_2, r_3, where $0 \leq \alpha \leq (n - 4)$, $0 \leq \beta \leq (n - 4)$ and $\alpha + \beta + 3 = n - 1$ (n is the accuracy level). Thus, altogether, there will be 8 combinations. Fig. 5.16 shows the errors in *CF* observed after three consecutive split-errors by setting $r_1 = 0$, $r_2 = 1$, $r_3 = 1$, for all values of starting-*CF*, and for all eight combinations of error-vectors. From the nature of the plot, it is apparent that it is very hard to predict for which error-vector the *CF*-error is maximized, even for a given combination of r-values. The maximum error depends on the *CF*-value from which the critical-split-section begins and also on the error-vector that is chosen (i.e., whether to proceed with the larger or the smaller daughter-droplet). Furthermore, the error-expression becomes increasingly complex when the number of split-errors becomes large. As an example, we perform experiments to study the fluctuations of the error in a particular target-*CF* for all combinations of error-vectors and showed the plot in Fig. 5.9.

From the above analysis and experimental results, it appears that the identification the vector for which the error in target-*CF* is maximized, is a difficult problem. In other words, it may not be possible to develop a procedure that will generate the worst error-vector without doing exhaustive analysis.

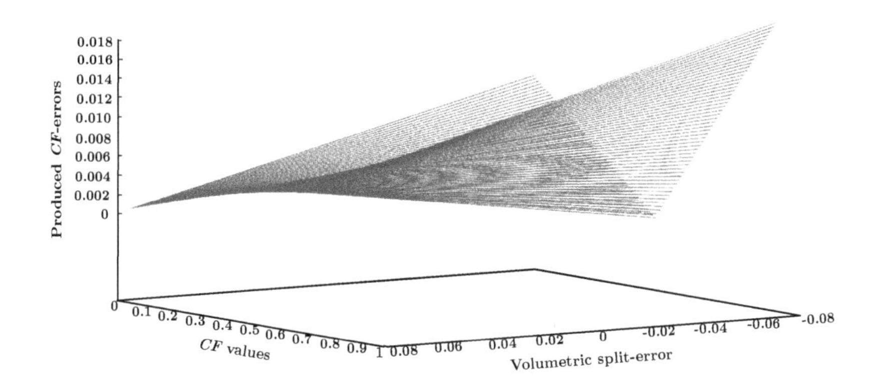

Figure 5.15: *CF*-error at the next mix-split step (for positive and negative single split-error): A different view

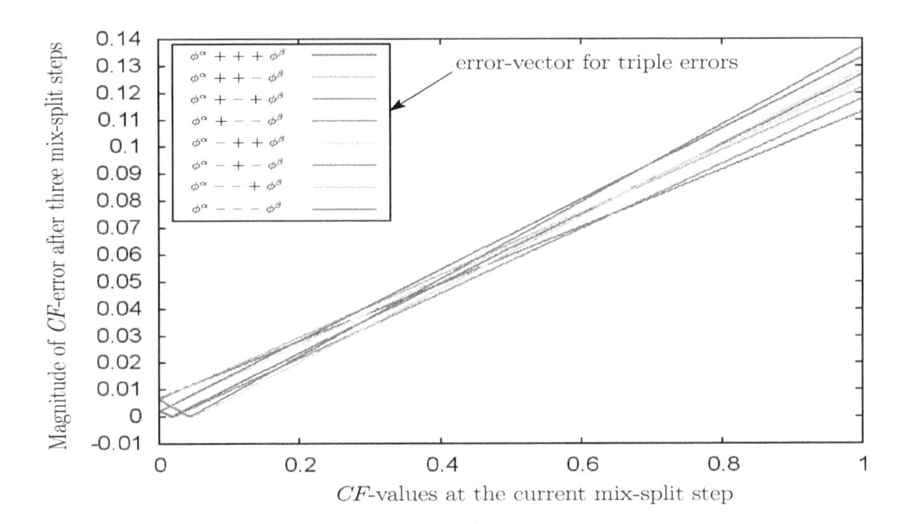

Figure 5.16: *CF*-error for triple split-errors

5.6 CONCLUSION

In this chapter, we have presented an analysis on the effect of single volumetric split-errors on a target-*CF* and observed that the maximum value of the *CF*-error in a target-droplet occurs for a negative split-error. The *CF*-error in a target-droplet increases with increasing magnitude of the split-error. Next, we perform various experiments to observe the effect of multiple split-errors on target-*CF*s and notice that the outcome may be affected by various combinations of erroneous droplets (smaller/larger) during the execution of mix-split operations. It is noticed that the *CF*-error in a target-droplet increases when the target-*CF* is affected by a large number of split-errors. We perform exhaustive simulation to identify the error-vector that maximizes the error in a given target-*CF*. Unfortunately, it appears that it is difficult to determine this vector using purely analytical methods.

6 Error-Oblivious Sample Preparation with DMFBs

In recent times, microfluidic biochips are being widely used for the efficient implementation of several biological protocols. Dilution and mixing of fluids are two major pre-processing tasks that are needed for most of them. In an automated assay, these two tasks are implemented on a digital microfluidic biochip (DMFB) as a sequence of droplet operations such as transport and mix-split steps. Most of the DMFB-based sample-preparation algorithms use (1:1) mixing model for performing mixing and splitting operation. However, droplet-splitting operation is very sensitive due to uncertain variabilities in fluidic operations on-chip, and thus often become erroneous, i.e., splitting may produce two unequal-volume droplets (unbalanced size). As a result, the concentration factor (*CF*) of the generated target droplets, i.e., the accuracy of sample preparation, i.e., is badly affected.

Unbalanced split operations obviously pose a significant threat to sample preparation. It is important to note that the reliability of many real-life biomedical applications (e.g., clinical diagnostics, drug design, or DNA analysis) depends on high-precision *CF*-values of the constituents in a fluid mixture where the outcome of the assay operations cannot be negotiated [59, 66, 88, 145]. Therefore, the resultant sample is rendered useless when the *CF*-error in the target droplets caused by unbalanced split operations exceeds certain error-tolerance limits.

Thus far, several error-recovery procedures based on re-execution (rollback) were reported to address this problem [3, 4, 58, 105, 106, 192]. In these methods, on detecting split-errors using on-chip sensors, the concerned portion of the assay is re-executed starting from a preceding non-erroneous checkpoint utilizing backup/copy-droplets stored therein. In the previous chapter, we have discussed a different error-recovery methodology named "roll-forward", which was first introduced Poddar *et al.* [122]. It cancels split-errors by executing the remaining portions of the task-graph in two parallel paths with the two erroneous daughter-droplets and then mixing the output droplets to neutralize the error.

However, all these aforementioned solutions suffer from several disadvantages. First of all, they require cyber-physical DMFBs with integrated sensors and a feedback controller for continuously monitoring the outcome of mix-split operations and sending the corresponding feedback signals to the controller in real-time. Apart from hardware overhead, they compromise badly with assay-completion time due to error-detection and subsequent re-execution. In many applications, fast error-recovery is needed where the reaction time lies in the order of milliseconds to seconds, e.g., in flash-chemistry (a powerful tool for drug discovery, clinical diagnosis, and novel material synthesis) [106, 187]. Furthermore, prior solutions assume that "successful re-execution is reliable" – which might not be true always, e.g., when a sufficient

DOI: 10.1201/9781003219651-6

number of back-up droplets are not available or when error re-occurs during the recovery cycle. Therefore, almost all previous approaches carry with them an overhead of assay completion time, reactant-cost, and also uncertainties in termination due to multiple recovery cycles.

In this chapter, we discuss a different solution methodology to address the problem of robust sample preparation (dilution) with DMFBs. More precisely, we present a new scheme called Error-Oblivious Sample Preparation (EOSP) that avoids sensing operations and re-execution of previous portions of the assay [124]. EOSP first investigates the effect of possible combinations of volumetric split-errors on the target-droplet. It is observed that some combinations of volumetric split-errors cancel out the unwanted effects and allow to achieve the error-tolerance, i.e., produce the target-droplet within given error-tolerance limits. These observations eventually lead to the development of a methodology that does not need feedback recovery loops driven by sensors. It modifies the task-sequencing graph that guarantees the production of target-droplets with desired accuracy. In other words, the procedure is oblivious to split-errors since there is no need for sensing operations and re-execution of assay-segments. EOSP aims to achieve a number of advantages for reliable sample preparation on DMFBs and has the following characteristics:

- it correctly produces the desired target-*CF* without the need for any re-execution operation;
- it is oblivious to volumetric split-errors, and hence, on-chip sensors are not required. It takes the same action whether or not any split-error has occurred;
- it can handle multiple split-errors in contrast to other approaches, which can handle a limited number of errors;
- it does not need any online resynthesis effort to correct the errors [95] as proposed in previous cyber-physical solutions, which may increase assay-completion time significantly [105];
- it does not require on-chip electrodes for storing back-up/copy-droplets Thus, it does not pose additional routing constraints to assay-droplets.

In the following, the underlying ideas, as well as the methodologies, are described in detail. Section 6.1 reviews the basics of sample preparation as well as split-errors that may occur during sample preparation. Section 6.2 illustrates the general concept of split errors using an example. In Section 6.3, we summarize their effects on target-droplets, which eventually lead to the development of the EOSP discussed in Section 6.5. Finally, in Section 6.6, experimental results are reported and the chapter is concluded in Section 6.7.

6.1 SAMPLE PREPARATION USING DMFBS

In this section, we discuss the basics of sample preparation with DMFBs and describe the errors that may occur while implementing dilution assays.

6.1.1 SAMPLE PREPARATION

Sample preparation is required in almost every assay because most of the collected samples cannot be utilized directly for downstream processing in a laboratory. For example, in the dilution problem, a raw sample fluid has to be diluted with a desired-CF ($0 < CF < 1$). More precisely, in order to produce the desired target-CF, we perform a sequence of (1:1) mix-split operations following certain algorithms [131,163], starting from the raw sample ($CF = 1$) and buffer fluid ($CF = 0$). The sequence of mix-split operations to be conducted can be envisaged as a directed acyclic graph (known as mix-split graph), where each node depicts the CF of a fluid droplet, and the incoming edges represent the droplets to be mixed. The outgoing edge represents a daughter-droplet (after a split operation) to be used for subsequent mix-split operations [131, 163].

Note that the depth of the mix-split graph/tree (also denoting the accuracy level n of a particular CF), is determined by a user-defined parameter " τ" (error-tolerance limit), where $0 < \tau < 1$. In order to bound the concentration error in the target-CF within the error-tolerance limit (τ) = $\frac{1}{2^{n+1}} = \frac{0.5}{2^n}$ [131, 163], each CF is represented as an n-bit binary fractional number $\frac{x}{2^n}$, where $x \in \mathbb{N}$, $0 \leq x \leq 2^n$, and $n \in \mathbb{N}$ [15]. However, for the sake of pictorial clarity, we will represent, each CF-value, for a given n, by its numerator only, i.e., we will write $CF = x$ in all figures, instead of $\frac{x}{2^n}$. The value of each CF is represented as the label of the corresponding node in the *mix-split graph/tree*.

Example 4 Fig. 6.1 shows the mix-split graph for preparing a diluted sample with $CF = \frac{25}{32}$ following the twoWayMix-algorithm [131, 163]. Note that edges of the mix–split graph determine the order of execution of each mix-split step.

6.1.2 ERRORS IN DMFB

During sample preparation in DMFBs, errors can frequently occur. We discuss below some common errors and their effects.

6.1.2.1 Dispensing error

Source droplets (i.e., sample or buffer droplets) are always admitted to a biochip from the input reservoirs placed around the boundary of the DMFB. Here, errors may occur leading to the emission of a droplet with incorrect volume. However, one can simply correct such errors by returning the erroneous droplet back to the source reservoir and re-dispensing again. Such a scheme requires overhead for sensing and re-dispensing, but saves costly reactants. Moreover, there exist other accurate droplet-emission mechanisms [27,30,46], which can be used to dispense precise volume of droplets from the reservoirs. Since reliable dispensing solutions are already available, we have not considered this type of error in this chapter.

6.1.2.2 Volumetric split error

In general, a split may be balanced, unbalanced, or residue-leaving (imperfect) as illustrated in Fig 1.6. In a balanced (unbalanced) split, two daughter-droplets

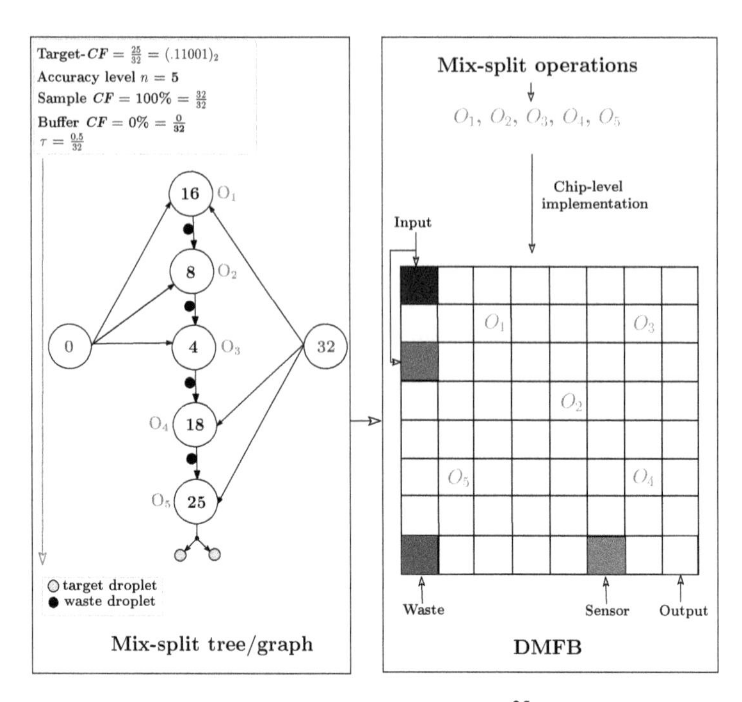

Figure 6.1: Mix-split graph for generating target-$CF = \frac{25}{32}$ for accuracy level 5. Only the numerator component of each CF-value is shown in the diagram; denominator = $2^5 = 32$ (with permission from IEEE [124])

are created with equal (unequal) volumes. Sometimes, a small residue is left on the middle electrode during a split operation while producing two equal-/unequal-volume droplets [75, 175]. Such operations not only make the volume of the split-droplets erroneous but also may cause cross-contamination because of the left-over residue [189].

Example 5 Consider the mix-split graph of Fig. 6.1 for preparing target-$CF = \frac{25}{32}$. Note that a single volumetric split-error of magnitude 7% in one of the operations in the mix-split graph of Fig. 6.2 (a) produces a CF-error = $\frac{0.2}{32}$. Moreover, if two or four volumetric split-errors of magnitude 7% occur, the CF-error rapidly grows to $\frac{0.4}{32}$ and $\frac{0.8}{32}$ as illustrated in Figs. 6.2 (b) and 6.2 (c), respectively.

6.1.2.3 Critical/non-critical set of errors

A set of volumetric split-errors is called *non-critical*, if the produced concentration error in the target-CF is $< \frac{0.5}{2^n}$, i.e., bounded by the inherent accuracy level determined by the choice of n, in spite of the occurrence of such multiple errors; otherwise they are termed as *critical* error. The sign of the error $+$ ($-$) indicates that the larger

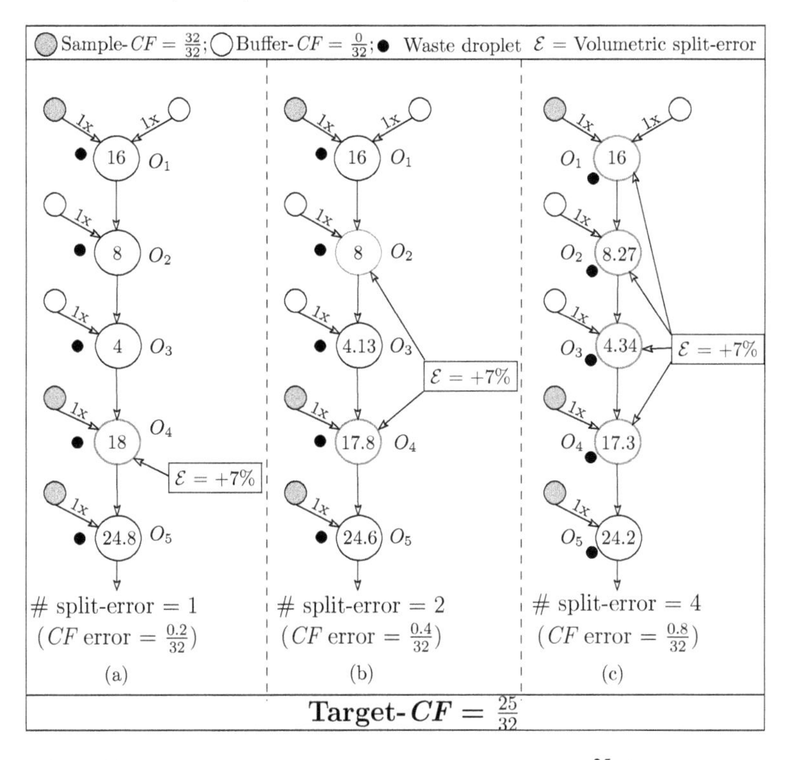

Figure 6.2: Effect of volumetric split-errors on target-$CF = \frac{25}{32}$ for accuracy level 5 (with permission from IEEE [124])

(smaller) daughter-droplet has been used in the following mixing-path. For example for accuracy level $n = 5$, the CF-error is bounded by $\frac{0.5}{32}$, and hence the errors in the mix-split operations $\{O_1, O_2, O_3, O_4\}$ represent a critical set of errors (since CF-error $= \frac{0.8}{32}$), whereas those in $\{O_4\}$ and $\{O_2, O_4\}$ are non-critical errors (Fig. 6.2). As observed from Example 5, the impact of the volumetric split-errors on the target-CF strictly depends on i) the magnitude and sign of the split-error (ε), and ii) the multiplicity of such errors. We assume that, whenever a volumetric split-error occurs, its magnitude is at most 7% [122, 129], and that they may occur in any mix-split step of the mixing-path.

6.1.3 SUMMARY OF PRIOR ART

Several sample-preparation algorithms for DMFBs have been reported, where each of them focuses on certain optimization objectives [100, 131, 133, 135, 163]; however, none of them addresses the impact of split-errors. In general DMFBs, sensor-driven error recovery techniques using pre-stored back-up droplets have been studied in [3, 4, 58, 105, 106, 192]. These approaches assume that the error-recovery operations are error-free [192], and other operations need to be suspended during the recovery process. In other error-recovery approaches [3, 106], actuation sequences are pre-stored in the memory (including those needed for error-recovery for anticipated errors) prior to the actual execution of the assay. By this, a limited number of errors (# of errors $<= 2$) can be handled – at the cost of memory overhead. In order to reduce the cost of error-recovery operations, certain portions of the assay may be re-executed [58, 105] using the previously stored back-up droplets. However, such on-chip back-up droplets require extra storage space and may create additional routing constraints to other active droplets.

In [4], a probabilistic approach to volumetric split-error correction has been proposed, where the split operation (involving the erroneous operation) is repeated up to a certain number of times to fix the error. Such a scheme requires additional time and the error may occur again after attempting the split operation a certain number of times. In roll-forward error-correcting method [122], erroneous droplets are used in a parallel mirror-path of the *task-graph* in order to correct the effect of volumetric split-errors on the target-*CF*, instead of discarding the erroneous droplets as waste.

Error recovery in Micro-Electrode-Dot-Array (MEDA)-based DMFBs has also been studied [94]. However, like all traditional rollback approaches, this method also increases the cost of error-recovery in both space and time dynamically using real-time sensor data. More recently, another hybrid error-recovery approach (rollback and droplet-volume regulation-based strategy) was proposed for MEDA-based DMFB chips [92]. However, it increases assay-completion time as error-detection needs to be performed after every mix-split step. The characteristics and scope of the EOSP method in contrast to prior approaches in the context of DMFBs are summarized in Table 6.1.

6.2 EOSP: MAIN IDEA

In EOSP, an alternative scheme for reliable sample preparation on DMFBs has been proposed. The main idea is not to use any sensor in order to check whether an error has occurred, and not to conduct re-computation to address the error. Instead, EOSP employs a straightforward scheme, which always guarantees the generation of the desired target-*CF* irrespective of possible errors. To this end, we exploit the fact that particular combinations of split-errors cancel out the unwanted effects.

More precisely, how a volumetric split-errors affects the generation of the desired target-*CF* depends on two factors, namely i) the positions of occurrences of such split-errors in the mix-split sequencing graph and ii) the volume of the selected daughter-droplet (smaller or larger one) that is used following such an erroneous mix-split step. Note that we have already discussed the effect of the former on target-*CF* in Section 5.3.2. The second factor can be analyzed in terms of error-vectors discussed in Section 5.4.

Table 6.1
Comparative features of EOSP [124] against prior art.

Method	Cyber-physical based?	Recovery mechanism	Recovery guaranteed?[1]	Consider worst-case error scenario?[2]	Reliable for multiple errors?	On-chip sensing needed?	Perform Re-synthesis?	Steps needed for recovery?
[192]	yes[5]	rollback[3]	no	no	no	yes	yes	indefinite
[3]	yes[5]	re-merge and re-split[4]	no	no	no	yes	yes	indefinite
[4]	yes[5]	rollback[3]	no	no	no	yes	yes	indefinite
[105]	yes[5]	rollback	no	no	no	yes	yes	indefinite
[106]	yes[5]	rollback	no	no	no	yes	yes	indefinite
[58]	yes[5]	rollback (dynamic)	no	no	no	yes	yes	indefinite
[122]	yes[5]	roll-forward	yes	no	yes	yes	yes	definite
[94]	yes[6]	rollback, re-merge and re-split[4]	no	no	no	yes	yes	indefinite
[92]	yes[6]	rollback and droplet-volume regulation	yes	no	yes	yes	yes	indefinite
EOSP [124]	no[5]	roll-forward	yes	yes	yes	no	no	definite

[1]split-error might reoccur during the re-execution phase; [2] based on proper selection of erroneous daughter-droplets (larger/smaller) in a sequence of mix-split steps; [3] the recovery sub-graph is assumed to be error-free during re-execution; [4] performs re-merge and re-split operations a fixed number of times; [5] DMFB based; [6] MEDA based.

6.2.1 ERROR-VECTOR (E)

Let us assume that k volumetric split-errors may occur, i.e., in k mix-split operations, two droplets with different volumes will result. Then, an error-vector $E = [+, -, \phi]^k$ denotes whether the smaller (denoted by $-$) or the larger (denoted by $+$) daughter-droplet is selected following these split-steps, and where ϕ denotes no-error. Thus,

3^k possible error-vectors are possible and the *mix-split graph* can be affected by any of them. Studies on the error-vector that maximizes the error in target-*CF* have been presented earlier in Chapter 5[1].

Example 6 Consider the mix-split graph/tree shown in Fig. 6.1 which, assuming an error-free execution, yields a droplet with $CF = \frac{25}{32}$. Furthermore, assume that a volumetric-split error occurs at the third and fourth mix-split step of this mix-split tree. Fig. 6.3 demonstrates the effect of the error-vectors on the target-*CF*.

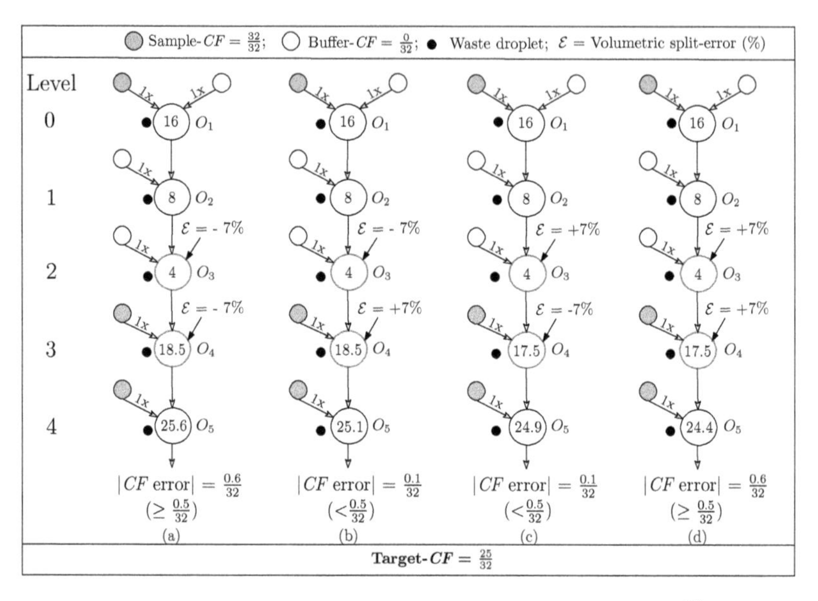

Figure 6.3: Effects of different types of error-vector on target-$CF = \frac{25}{32}$ for accuracy level 5, involving two erroneous steps (with permission from IEEE [124])

We also observe that some subsets of errors mutually cancel out their effects and, hence, do not have an impact on the target-*CF* (*non-critical set of errors*). On the other hand, some subsets of error may affect the target-*CF* badly (critical errors), for which we need to correct them by modifying the corresponding mix-split steps using a roll-forward approach [122] or other methods.

Example 7 Figs. 6.4 (a) and (b) show the effect of the error-vectors $[+,-,-,+]$, $[+,+,+,+]$ on target-$CF = \frac{25}{32}$ for accuracy level 5. Out of $3^4 - 1 = 80$ possible error combinations, it is observed by exhaustive simulation that the error in the target-*CF* becomes maximum for vector. $[+,+,+,+]$. However, if we consider single split-errors with the same sign, then each individual step $\{\{O_1\}, \{O_2\}, \{O_3\}, \{O_4\}\}$ becomes non-critical. On the other hand, by simulating all 80 errors, we observe that in

[1]Note that we consider errors at all mix-split steps. Thus, the number of error-vectors grows exponentially with k. However, we cope with the resulting complexity with a dedicated scheme for determining the error-vector as described later in Section 6.3.2.

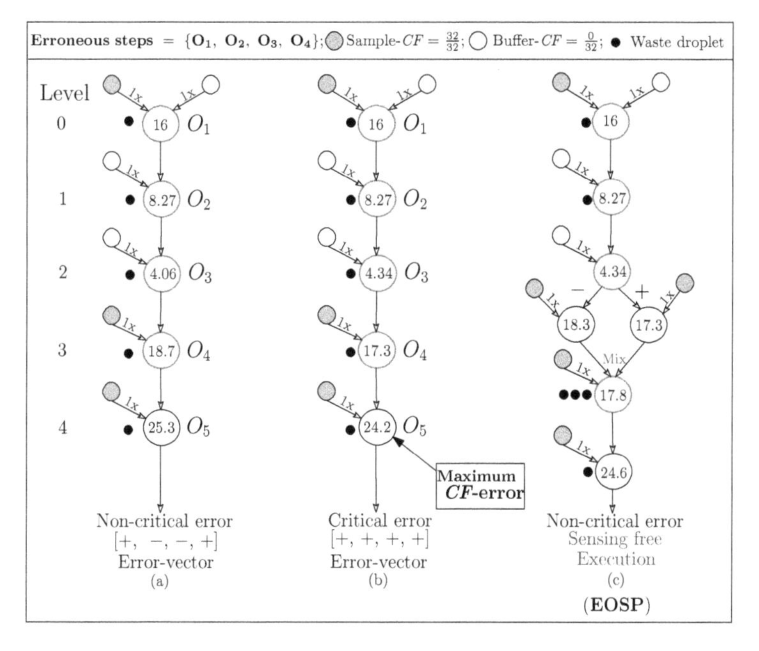

Figure 6.4: General idea of EOSP (with permission from IEEE [124])

the presence of all multiple split-errors, $\{O_1, O_2, O_4\}$ turns out to be the maximum-size non-critical subset of errors for this target-*CF*. That is, the effects of these mix-split errors cancel out each other – regardless of whether an error occurs or not. Based on this fact, the original *mix-split tree* is changed only at the output of Step O_3 as shown in Fig. 6.4(c). Conducting the mix-split steps according to this graph eventually yields the target-*CF* within the error-tolerance limit even in the presence of errors (and without the need for any sensing or re-computation). Note that in the modified graph, error occurring in Step O_3, if any, will cancel out at Step O_4, because both the two erroneous droplets (one larger and one smaller), are allowed to execute identical actions before merger at O_4 [122]. Error-oblivious sample preparation (EOSP) has been developed based on this idea and it generates the desired *CF* regardless of any split-error. Details of the method are presented in Section 6.3 and in Section 6.5, and Section 6.6 reports experimental results.

6.3 EFFECT OF ERRORS

In this section, we present a rigorous analysis that reveals the role of volumetric split-errors on a *CF* during sample preparation. We have discussed the effect of multiple volumetric split-errors on a target-*CF* in Section 5.3.2. We will now revisit the notion of critical and non-critical sets of errors in more detail. Then, a rule of thumb is stated in the next subsection which allows to maximize the error in a target-*CF*. The EOSP methodology is described in Section 6.5.

6.3.1 CRITICAL AND NON-CRITICAL SET OF ERRORS

We consider *multiple volumetric split-errors* (which can appear in any combination on the *mix-split graph*) in our error model. We measure their effects on the target-*CF* using Expression (5.1) and Expression (5.2) and based on that, we classify them as being *critical* or *non-critical*. Clearly, the non-critical set of errors can easily be ignored as there will be no significant change in the output-*CF*. Otherwise, corrective measures are required to restore the desired target-*CF* for the critical set of errors.

Table 6.2
Effect of volumetric split-errors on target-*CF* = $\frac{25}{128}$ for accuracy level n=7.

Erroneous steps*	*Produced* CF×128	CF *error (%)*	Non-critical set?
$\{O_1\}$	25.04	0.17%	Yes
$\{O_1, O_3\}$	25.14	0.56%	Yes
$\{O_2, O_3, O_4\}$	25.38	1.52%	Yes
$\{O_1, O_2, O_3, O_4\}$	25.43	1.72%	Yes
$\{O_1, O_2, O_3, O_4, O_6\}$	26.28	5.11%	No
$\{O_1, O_2, O_3, O_4, O_5, O_6\}$	27.59	10.36%	No

*for error-vector $\{O_1, O_2, O_3, O_4, O_5, O_6\} \rightarrow [+,+,+,+,+,+]$.

Example 8 Consider multiple volumetric split errors in a mix-split graph which aims for generating target-*CF* = $\frac{25}{128}$ with accuracy level $n = 7$. Note that there exist $\{3^{n'-1} - 1\}$ possible multiple split-errors of a mix-split graph containing n' mix-split operations. Table 6.2 demonstrates the effect of volumetric split-errors to the target-*CF* for some representative test cases with 7% volumetric split-errors.

We carried out additional experiments for counting the number of affected target-*CFs* by increasing the number of errors for the odd target-*CFs* of accuracy level 7, and report the results as barchart in Fig. 6.5. We conduct the same experiment for different percentages of volumetric split-errors $(\varepsilon) = 3\%$, 5% and 7%. Note that it is possible to insert l errors at different mix-split steps in $\binom{n'}{l}$ ways. In our experiment, we explore all these possibilities and take the average of the outcomes of a target-*CF* for given l. Fig. 6.5 reveals that the count of target-*CF*s becomes null for single volumetric split-error of magnitude 3% and 5%. However, the population of the affected target-*CF*s increases gradually with the increase in the number of volumetric split-errors (4-6) as well as in the quantum of error-percentage from 3% to 7%.

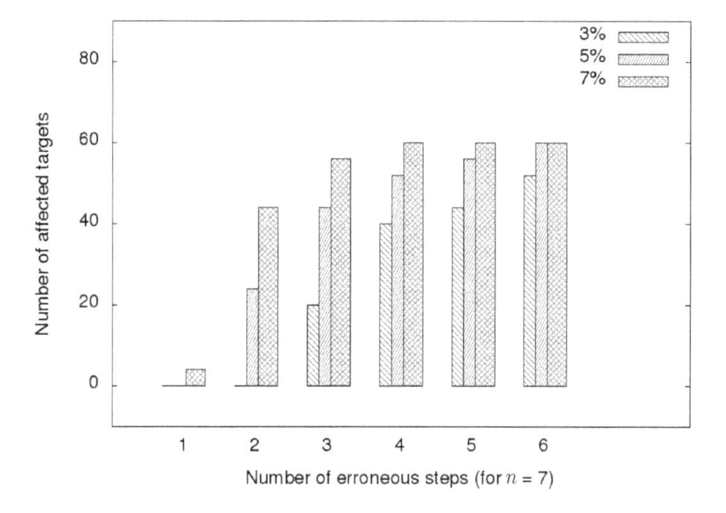

Figure 6.5: Number of affected targets for different split-errors for all odd target-*CFs* with accuracy level $n = 7$ (with permission from IEEE [124])

6.3.2 EFFECT OF MULTIPLE ERRORS ON TARGET-CFS

As discussed in Section 5.3.2 a volumetric split-error produces two unequal droplets of volume $(1+\varepsilon)$ and $(1-\varepsilon)$ for some value of ε, $0 \leq \varepsilon < 1$. In the case of a mix-split operation involving a split-error, it is very hard to predict whether the smaller or the larger one has been forwarded to the next mix-split operation in the absence of any error-detection mechanism. The situation becomes more complex when several mix-split operations suffer from split-error. It is also interesting to determine the set of split-operations that yields the maximum error at the target-*CF*. The corresponding error-vector is likely to contain a large number of critical-steps in the mix-split sequence.

Example 9 Consider the generation of target-$CF = \frac{25}{128}$ with accuracy level 7. Let us assume that split-errors appear in each split operation of the mix-split graph (except the final one, and hence it is not explicitly shown as an element in the error-vector). For the error-vector $[+,+,-,+,+,+]$, the error in the target-CF ($\frac{27.70}{128}$) becomes maximum (see Table 6.3). Searching for the vector requires $O(N)$ time, where N is number of entries in the search table. Since $N = 3^{n'-1} - 1$, the search time increases exponentially with n'.

Note the error-vector that causes maximum error in the target-*CF* represents the worst-case scenario. However, for many other error-vectors, the deviation in the target-*CF* may also be far outside the acceptable limit. Hence, for achieving fault-tolerance, the modified mix-split graph should be made resilient to each and every such potential error-vector. Later in this section, we will describe a procedure for

Table 6.3
Effect of various error-vector on target-*CF* = $\frac{25}{128}$ for accuracy n=7.

Serial no.	Error-vector	Produced $CF \times 128$	CF-error (%)
1	$[+,+,+,+,+,+]$	27.59	10.36%
2	$[+,+,+,+,+,-]$	25.78	7.12%
3	$[+,+,+,+,-,+]$	24.91	4.36%
4	$[+,+,+,+,-,-]$	23.22	7.12%
.	.	.	
.	$[+,+,-,+,+,+]$	27.70	10.80%
.	$[-,-,+,-,-,-]$	22.31	10.76%
.	.	.	
61	$[-,-,-,-,+,+]$	26.73	6.92%
62	$[-,-,-,-,+,-]$	24.92	0.32%
63	$[-,-,-,-,-,+]$	24.05	3.80%
64	$[-,-,-,-,-,-]$	22.37	10.52%

graph-modification that starts from an error-vector, which involves a large number of critical steps for the target-*CF*. We may start from the maximum-error-vector obtained by exhaustive simulation. However, in order to avoid the search complexity, we present below an empirical rule-based procedure that produces an error-vector, which causes near-maximum *CF*-error at the target. To this end, one needs to scan the nodes (*CF*-values) of the mix-split graph starting from the root to the leaf node (target node). More precisely, during this computation, the following two sets are determined:

- H_{cf}: the set of all intermediate *CF*s that have a larger value than the target-*CF* and
- L_{cf} : the set of all intermediate *CF*s that have lower value than the target-*CF*.

Following two cases may now occur:

Case I. Let us assume that the target-*CF* lies within the interval (0, 2^{n-1}-1], and $\{C_i, C_j, C_k\}$ are the intermediate *CFs*, $C_i \in H_{cf}$ and $C_j \in L_{cf}$, i.e., $C_i > C_t$ and $C_j < C_t$. Now, a larger-volume droplet is selected in an erroneous node C_i ($CF = C_i$) for the edge representing mix-split operations from C_i to C_k in the *mix-split graph*, i.e., ($C_i \rightarrow C_k$). However, a smaller-volume droplet is selected at an erroneous node C_j for the edge $C_j \rightarrow C_k$.

Case II. When the target-$CF \in [2^{n-1} + 1, 2^n$ - 1) with $C_i \in H_{cf}$ and $C_j \in L_{cf}$, the larger-droplet is selected in an erroneous node C_j during the mix-split operation ($C_j \rightarrow C_k$), whereas the smaller-droplet is selected for the mix-split operations ($C_i \rightarrow C_k$).

We validate the rule stated above by performing a number of experiments. We run experiments for each of the target-*CF*s with various accuracy levels, and generate the error-vector. We inject volumetric split-errors at each mix-split operation of the mix-split graph according to the error-vector governed by the above empirical Rule. Exhaustive simulation shows that it indeed produces the maximum *CF*-error in most of the cases barring only a few instances. Fig. 6.6 shows the selection of the error-vector $[+,+,-,+,+,+]$ for target-$CF = \frac{25}{128}$. The *CF*-values highlighted with red (black) indicates the deviated (expected) values due to the presence (absence) of errors in the mix-split operations. The effect of selecting an erroneous droplet (+/−) on the expected *CF* for each mix-split step of the mix-split path is also shown in red font.

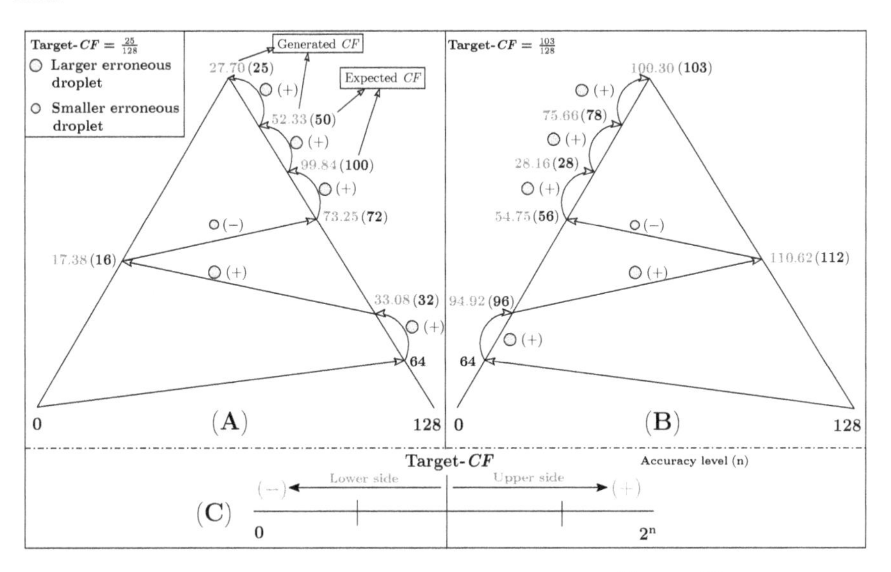

Figure 6.6: Selection of erroneous droplets for target-*CFs* $\frac{25}{128}$ and $\frac{103}{128}$ with accuracy level $n = 7$ (with permission from IEEE [124])

We further performed experiments for determining the critical and non-critical set of errors for various error-vectors. It has been noticed that the non-critical set $\{O_1, O_2, O_3, O_4\}$ in Table 6.2 becomes critical for the error-vector $[+,+,-,+]$. We have performed rigorous theoretical analyses and further experiments to study the properties of maximum *CF*-error in a target-*CF*, the details of which have been presented in Chapter 5 and in [123]. Note that given a target-*CF*, Algorithm 3, described later, selects an error-vector based on the earlier empirical rule.

6.4 BASELINE APPROACH TO ERROR-OBLIVIOUSNESS

We present below a baseline approach for canceling the effects of multiple split-errors regardless of their criticality. The idea is to construct a complete binary mix-split tree (where each node excepting the leaves has degree two, and all leaf nodes are at depth n, where n is the accuracy level). For simplicity, we prove the theorem for $n = 4$. However, the argument can be extended for any general value of n. Let us consider the problem for generating a target-$CF = C_t$ with accuracy level $n = 4$. Let $\{O_1, O_2, O_3, O_4\}$ be the sequence of (1:1) mix-split operations which needs to be performed for generating the target-CF. The complete dilution tree for producing $2^4 = 16$ target-droplets with $CF = C_t$ is shown in Fig. 6.7. In this tree, all nodes present in a particular depth are identical and a node at i^{th} depth corresponds to the $(i+1)^{th}$ mix-split operation O_{i+1} for the target-$CF = C_t$. For example, nodes O_2^L and O_2^R in depth 1 are identical, and they represent the second mix-split operation O_2 for the target-$CF = C_t$. The target-droplets $(d_1 - d_{16})$ appear as leaf node of the dilution tree, and in the normal condition (in the absence of volumetric errors) each of them will have the same CF-value. We can now prove the following result.

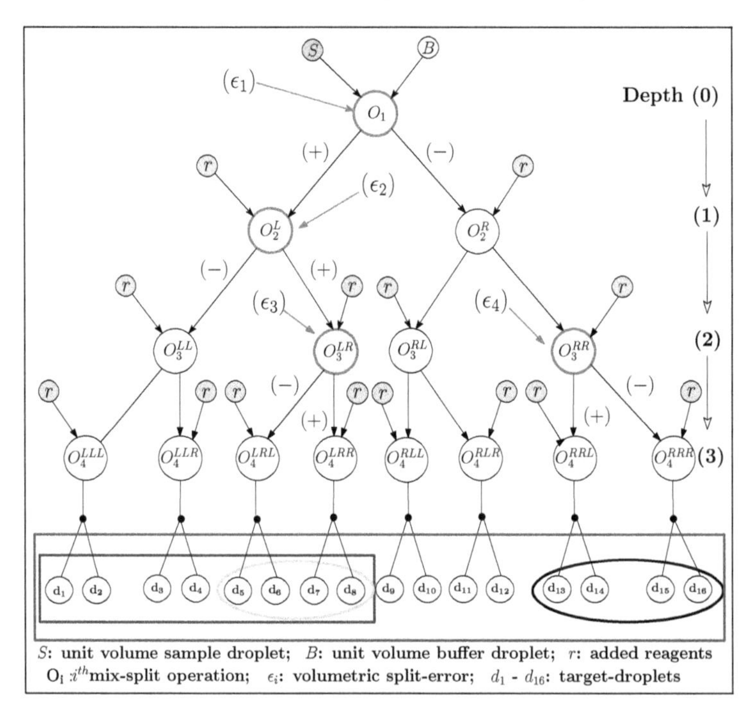

Figure 6.7: Error-free sample preparation with multiple split-errors (n = 4) (with permission from IEEE [124])

Theorem 2 If the target-droplets $\{d_1, d_2, \ldots, d_{2^n}\}$ produced by the complete dilution tree (Fig. 6.7) are mixed together at the end, without discarding any intermediate droplets, then the CF of the resulting mixture becomes exactly C_t even if volumetric split-errors occur in the mixing tree.

Proof Without loss of generality, let us assume that while executing the operations of the dilution tree (Fig. 6.7), mix-split operations O_1, O_2^L, O_3^{LR}, and O_3^{RR} (red colored nodes) are affected by the split-errors $\varepsilon_1, \varepsilon_2, \varepsilon_3$, and ε_4. Note that an erroneous split operation produces two unbalanced daughter-droplets, one with a larger volume (+) and another with a smaller volume (-), as shown in Fig. 6.7. However, the effect of an erroneous split-operation to the target-*CF* is canceled-out by mixing the target-droplets (produced by the two erroneous droplets) together at the end, when the error occurs at their nearest-common-ancestor mix-split node (step) [122]. For example, if we mix target-droplets (d_5, d_6) and (d_7, d_8) together at the end, the effect of splitting-error ε_3 (occurred at the nearest-common-ancestor mix-split node O_3^{LR}) is canceled-out. Although mixing of (d_5, d_6) and (d_7, d_8) removes the effect of ε_3, it does not remove the effect of ε_2. However, the effect of ε_2 on the target-droplet is canceled when droplets (d_1, d_2, d_3, d_4) and (d_5, d_6, d_7, d_8) are mixed together at the end. Thus, the overall effects of ε_2 and ε_3 on target-droplets d_1 to d_8 are removed when they are mixed. Similarly, we can argue that the effect of ε_4 is corrected at the end by mixing target-droplets (d_{13}, d_{14}) and (d_{15}, d_{16}). Finally, mixing of target-droplets $(d_1$-$d_8)$ and $(d_9$-$d_{16})$ not only removes the effect of split-error ε_1 but also cancels the effects of $\varepsilon_1, \varepsilon_2$ and ε_3 at the end. Hence, mixing of all target-droplets produces a final solution with $CF = C_t$ regardless of the number of volumetric split-errors in the mixing tree. This argument can be generalized for any complete dilution tree.

Therefore, it is possible to generate a target-*CF* correctly from the complete dilution tree regardless of the number of volumetric split-errors. However, this baseline approach not only increases sample-preparation time but also increases reactant-cost significantly, as the dilution tree contains 2^n-1 mix-split nodes. In order overcome this issue, we create a small-size[2] graph (called *error-tolerant graph*) for producing the target-*CF* within the error-tolerance limit. We may start with selecting an arbitrary error-vector (E), and then prune/modify the complete binary tree by mapping each critical and non-critical error in E to the corresponding node of the complete dilution tree for the given target-*CF*. However, it is preferable to start from an error-vector that causes the maximum-error in target-*CF* (obtained by exhaustive search) or from the one produced by the empirical rule, which produces an error-vector that "almost" maximizes the error in the target-*CF*. Such an initial choice usually speeds up the construction of the error-tolerant graph. The above method clearly works if we start from any other error-vector, but in that case, search complexity for finding the final critical steps may increase in the later phase. We will also show that the proposed method reduces sample-preparation time as well as reduces reactant-cost noticeably compared to the baseline approach.

6.5 RESULTING METHODOLOGY

The working principle of EOSP for generating target-$CF = C_t$ within the user specific error-tolerance bound (τ) on a digital microfluidic platform is as follows [124].

[2]reduced # of mix-split steps

- Construct the mix-split path for generating a target-*CF* using the *twoWayMix* algorithm [163] and the corresponding complete dilution tree (T) [133].
- Depending on the maximum allowable split error (ε), determine the error-vector by the empirical rule as described in Section 6.3.2; this produces a near-maximum error in the target-*CF*.
- Determine the initial set of "critical-steps" (considering only one split-error at a time) with the sign matching with that of the error-vector obtained in the previous step. Each of these critical steps will have to be corrected later in the mix-split graph using roll-forward action. Let nc_{steps} denote the set of remaining steps, which are individually non-critical for this error-vector.
- Now we need to check the criticality of all mix-split steps belonging to nc_{steps} from a collective viewpoint (i.e., in the presence of simultaneous occurrences of two or more such errors). In order to test this, we assume that the critical steps found so far have been corrected; next consider all $3^{|nc_{steps}|-1} - 1$ combinations (+, -, ϕ (no-error)) of error-vectors. A k-size subset of nc_{steps} is said to be *finally non-critical* if for each assignment of + and - on these k-positions, the corresponding multiple split-error, when inserted, causes the target-*CF* to lie within the allowable limit. We begin the search starting from higher-to-lower order (i.e., first with the combinations that comprises no ϕ, and later increasing their occurrences), and exit as soon as a non-critical subset of steps is identified. This will be a maximal-size non-critical subset of steps (nc_{max}) on the reaction-path. The remaining steps belonging to $\{nc_{steps} \setminus nc_{max}\}$ are marked critical, and are added to the initial set of critical steps.
- For each mix-split operation at i^{th} depth of T, starting from the root (depth 0),
    - if $i \notin nc_{max}$
        * retain all the nodes at depth (i+1), which have parent-child relationship with the existing nodes at depth i.
    - if $i \in nc_{max}$
        * keep a single node at depth (i+1) of T and delete all other nodes.[3]

We present below two algorithms: Algorithm 3 generates an error-vector based on the empirical rule discussed earlier in Section 6.3.2; Algorithm 4 returns a maximal-size non-critical subset of mix-split steps based on the working principle discussed in this section. Given the desired target-*CF*, the required number of mix-split steps (l), and the maximum allowable volumetric split-error, Algorithm 3 calculates *CF* of the intermediate fluid following each mix-split operation (Line 1) and based on that it determines the corresponding error-vector (Lines 3-14).

[3]during bioasssay execution, all the droplets generated at the ith mix-split step are mixed together from which only a single droplet is extracted for subsequent use.

Algorithm 3: Error-vector based on empirical rule.

Input: Target-$CF = C_t$, accuracy level n, size of the mixing path l, maximum allowable split-error ε.
Output: Error-vector E_v that causes a near-maximum error in the target-$CF = C_t$.

```
1  Let I_cfs = [Cf_1, Cf_2, Cf_3, ..., Cf_l] be the CFs corresponding to the sequence of intermediate mix-split
     operations for the target-CF;
2  Let E_v = [] // used for storing the error-vector;
3  for (i = 0; i < |l|; i = i+1) do
4      //for each mix-split step;
5      if C_t ∈ [1, 2^{n-1} − 1] then
6          if I_cfs[i] > C_t then
7              E_v.append(+ε)
8          else
9              E_v.append(−ε)
10     if C_t ∈ [2^{n-1} + 1, 2^n − 1] then
11         if I_cfs[i] > C_t then
12             E_v.append(−ε)
13         else
14             E_v.append(+ε)
15 return E_v
```

Algorithm 5 (Line 3-5) takes care of the case when C_t becomes $\frac{2^{n-1}}{2^n}$.

Example 10 Consider target-$CF = \frac{25}{128}$ for accuracy level 7 and allowable split-error 7%. First, it calculates the CF-value after each mix-split operation while generating the target-CF: $I_{cfs} = \{64, 32, 16, 72, 100, 50\}$, using the twoWayMix method. Next, by scanning the intermediate CFs (I_{cfs}) from left-to-right, Algorithm 3 returns the error-vector [+, +, -, +, +, +] for the target-CF [see Fig. 6.6].

Algorithm 4 returns a maximal-size non-critical set of errors depending on the target-CF, accuracy n, the error-vector returned by the Algorithm 3, and the error-tolerance limit τ. Procedure Gen_{CF} returns the generated CF-value based on the positions of split-errors and the error-signs. For example, if we insert non-zero error in all six mix-split steps (excepting the last one), there will be 63 possible error-vectors for the target-CF $\frac{25}{128}$ for accuracy level = 7. The generated CF-errors corresponding to these error-vectors are shown along Y-axis in Fig. 6.8 for split-error ε = .07. In the X-axis, the error-vectors are sequenced following a gray-code (assuming + $\rightarrow$0, and $-$ $\rightarrow$1). Note that the error in the target-CF exceeds the safe limits for a number of error-vectors. Algorithm 4 returns $\{O_1, O_2, O_3\}$ as the maximal-size non-critical steps. Other non-critical maximal sets also exist such as $\{O_1, O_2, O_4\}$ corresponding to the error-vector [+, +, -, +, +, +].

Algorithm 5 creates the error-tolerant graph by pruning and modifying the complete binary mix-split tree based on the non-critical set of errors (nc_{max}) produced by Algorithm 4. In Line 9, it determines maximal-size non-critical set of errors for the target-CF. Next, for each increasing depth of the complete dilution tree T, Algorithm 5 checks whether or not, the split operation performed at depth i belongs to the critical set of errors. If not, a single node at $(i+1)^{th}$ depth of T is retained and all other nodes at depth $(i+1)$ are removed. It means that during execution, only one daughter-droplet will be used for performing the mix-split operation at the next step.

Algorithm 4: Find maximal non-critical (nc_{max}) steps.

```
Input: Target-CF = C_t, accuracy level n, error-vector e_v and error tolerance limit = τ.
Output: Maximal-size non-critical set of errors (nc_max) = {O_1, O_2, O_3, ... O_k}, where 1 ≤ k ≤ n.
1  Express C_t as n-bit binary number (bin_Ct) x/2^n;
2  Remove the low-order zero bits from bin_Ct;
3  Let l = length(bin_Ct) - 1;
4  Let S = [1, ..., l];
5  Let NC_steps = [] // used for storing all possible non-critical steps;
6  Let nc_max = [] // used for storing max-length non-critical steps;
7  for (i = 0; i < |S|; i = i + 1) do
8      //for each position of S;
9      C = Gen_CF(S[i], e_v[i], l, bin_Ct)
10     if |C_t - C| / 2^n < τ then
11         NC_steps.append(S[i]) // S[i] is merged as non-critical step
12 Let S be the all possible subsets of nc_steps, except the null set; // for finding the remaining critical-steps w.r.t
   other error-vectors;
13 for (i = |S| - 1; i >= 0; i = i - 1) do
14     //for each subset of S;
15     Let E_v = all error-vectors of length S[i];
16     flag = 0;
17     for (j = 0; j < |E_v|; j = j + 1) do
18         //for each error-vector of E_v;
19         C = Gen_CF(S[i], E_v[j], l, bin_Ct);
20         if |C_t - C| / 2^n >= τ then
21             flag = 1;
22             break;
23     if flag == 0 then
24         nc_max = S[i] // max-length non-critical set obtained;
25         break;
26 return nc_max
1  Procedure Gen_CF (E_p, E_v, l, bin_Ct)
3      P_1 = 1/2, Q_1 = 1, C_1 = 1/2, V_1 = 1;
4      for (j = 2; j <= l + 1; j = j + 1) do
5          ε = 0;
6          if (j - 1) ∈ E_p then
7              ε = E_v[E_p.index(j - 1)]
8          end
9          P_j = P_{j-1} × (1 + ε) + 2^{j-2} × bin_Ct[j];
10         Q_j = Q_{j-1} × (1 + ε) + 2^{j-2};
11         C_j = P_j / Q_j, V_j = Q_j / 2^{j-1};
12         /* The subscript of P_j, Q_j and C_j changes according to the value of the for-loop j and P_1, Q_1 and
           C_1 are initialized in line 3;
13     end
15     return C_j
```

On the other hand, if the mix-split step at the ith depth belongs to a critical set of errors, then all nodes at $(i+1)^{th}$ depth are retained in T, and in this case, all droplets generated at i^{th} depth are used to perform the mix-split operations in the next level.

It is easy to prove that the modified mix-split graph constructed by Algorithm 5 will tolerate multiple split-errors of any sign occurring at any step. Note that the use of "maximum-error-vector" to construct an error-tolerant graph is a sufficient condition but *not necessary* for ensuring error-tolerance. We justify below why the EOSP method for producing an error-tolerant graph correctly works, regardless of the initial choice of the error-vector.

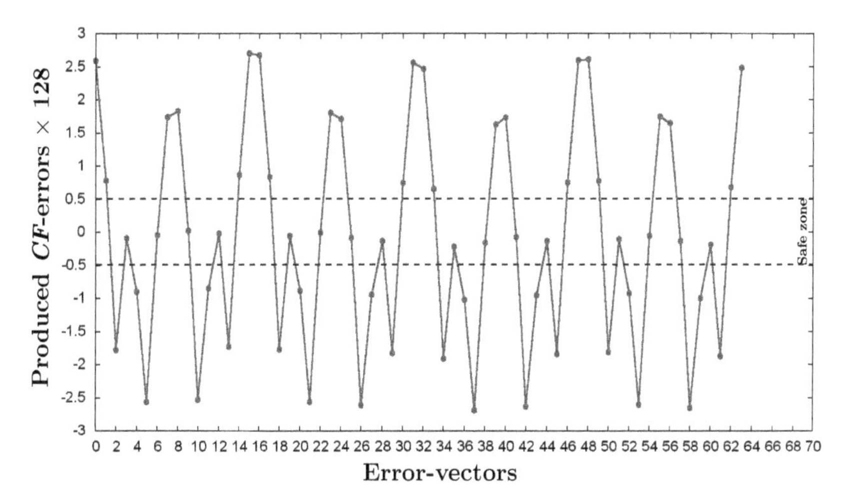

Figure 6.8: *CF*-errors corresponding to various error-vectors (with permission from IEEE [124])

- Step 1: Given a target-*CF*, we may start from any arbitrary error-vector (E, $|E| = n$ - 1) or the one generated by the empirical rule (Algorithm 3), or with the vector that produces the maximum-error (obtained by exhaustive simulation, by injecting maximum allowable split-error in each mix-split step, concurrently);

Algorithm 5: Error-tolerant dilution tree (T).

Input: Target-$CF = C_t$, maximum allowable split-error ε, error-tolerance limit τ, accuracy level n.
Output: Error-tolerant dilution (T) tree for the target-$CF = C_t$.

```
1  Let O = [O_1, O_2, O_3, ..., O_k] be the set of mix-split operations need to be performed for generating the
     target-CF = C_t, 1 <= k <= n;
2  Let T be the complete dilution tree [133];
3  if length(O)==1 then
4  |   //execute when target-CF = (2^{n-1})/(2^n) = 1/2;
5  |_  return T
6  else
7  |   //for other CFs;
8  |   Let E_v be the error-vector returned by Algorithm 3 ;
9  |   Let nc_max be the maximal-size non-critical set of error returned by the Algorithm 4;
10 |   Let CR_ops =[c_1, c_2, c_3, ..., c_q] be the critical set of errors, 0 <= q <= k;
11 |   for (i = 0; i < |O|; i = i+1) do
12 |   |   //for each depth of the dilution tree (T), starting from 0;
13 |   |   if i ∈ nc_max then
14 |   |   |_  Keep a signle node at depth (i+1) of T and delete all other nodes.
15 |   |   else
16 |   |   |   Retain all the nodes at depth (i+1) in T, which have parent-child relationship with the
   |   |_  |_    existing nodes at depth i.
17 |_  return T
```

- Step 2: We simulate E on the complete dilution tree T considering only one error at a time, and determine the set of critical steps, if any, by checking the observed target-CF at the output of T; these steps constitute the initial set of critical and non-critical steps;
- Step 3: We then consider the bit-positions in E that correspond to non-critical splits for the particular error-vector E, and let there be k such bits, $k < n$ -1, where n is the accuracy level of CF. There will be altogether 3^k possible error-vectors consisting of "+" error, "-" error, and "ϕ" (no error) for these positions. A subset of k' ($k' \leq k$) bit-positions is called non-critical if (k - k') bit-positions are set to ϕ, and if the final target-CF lies within allowable limits for every possible combinations of "+" and "-" errors, on k' bit-positions, while simulated on the dilution tree, T. We repeat this experiment for all positions of ϕ-bits. We run simulation starting from k' = k, through 0, and find the largest k' (k'_{max}), and their positions to identify a maximal-size subset of non-critical steps (nc_{max}) for the maximum allowable split-error (ε). We then set the mix-split steps corresponding to the remaining (k - k'_{max}) bits, to critical steps.
- Step 4: For each non-critical node, we keep only a single node at the next depth of T by deleting all other nodes, to create the error-tolerant graph G. In other words, for the new set of critical-nodes, roll-forward sub-graphs are retained in T for creating the error-tolerant graph.

It is easy to prove that G will tolerate multiple split-errors of any sign occurring at mix-split steps. Needless to say, if any error occurs at the critical nodes, they are canceled because of roll-forward mechanism. For the remaining non-critical nodes, no combination of error-vectors can produce a critical error at the target because they have been exhaustively checked while determining the non-critical set of steps (in Step 3 of the method stated above).

For example, a complete dilution tree (T) for accuracy level n = 5 is shown in Fig. 6.9, where red circles represent critical operations, green ones denote non-critical operations, dotted lines represent the virtual path, and solid lines represent the actual execution-paths in the dilution tree. Note that the baseline approach generates all 32 target-droplets by executing T for ensuring the correctness of the target-CF. In contrast, EOSP produces only 4 target-droplets. In this particular case, it only executes the paths shown by solid lines in T as shown in Fig. 6.9. However, for the ease of visualization, the final error-tolerant dilution tree, produced by Algorithm 5 is shown in Fig. 6.10. In this case, the *critical errors* are $\{O_2, O_4\}$, and *non-critical errors* are $\{O_1, O_3, O_5\}$. The values of target-CF for all error-vectors for target-CF = $\frac{25}{128}$ are shown in Fig. 6.11. It can be seen that, for all combinations of split-errors inserted in the graph, the resulting target-CF always lies within the allowable limits $\pm\frac{0.5}{128}$. Note that the empirical-rule-based Algorithm 3 is used for two reasons: (i) it gives the target-CF-error that is very close to the maximum, and (ii) it is likely to have a large number initial critical steps with respect of single split-errors, it obviates the need for time-consuming exhaustive simulation for all full-length error-vectors, which will be $O(3^{|E|})$= $O(3^n)$.

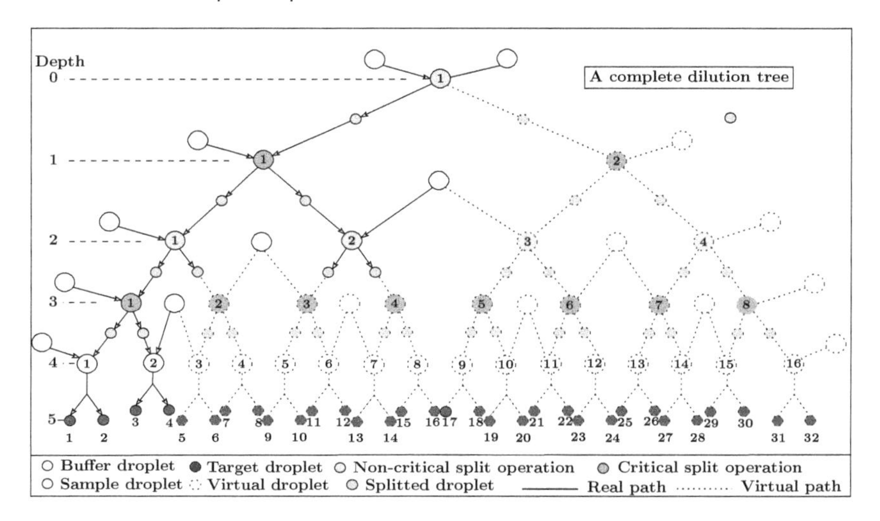

Figure 6.9: A complete dilution tree for generating a target-*CF* for accuracy level 5 (with permission from IEEE [124])

Handling residue-leaving (imperfect) splitting: As illustrated earlier in Fig. 1.6, some split operations may leave a small-size residual droplet in addition to producing two daughter-droplets. Without using any sensors, such errors can be handled by incurring some routing overhead: after each splitting, one of the daughter droplets

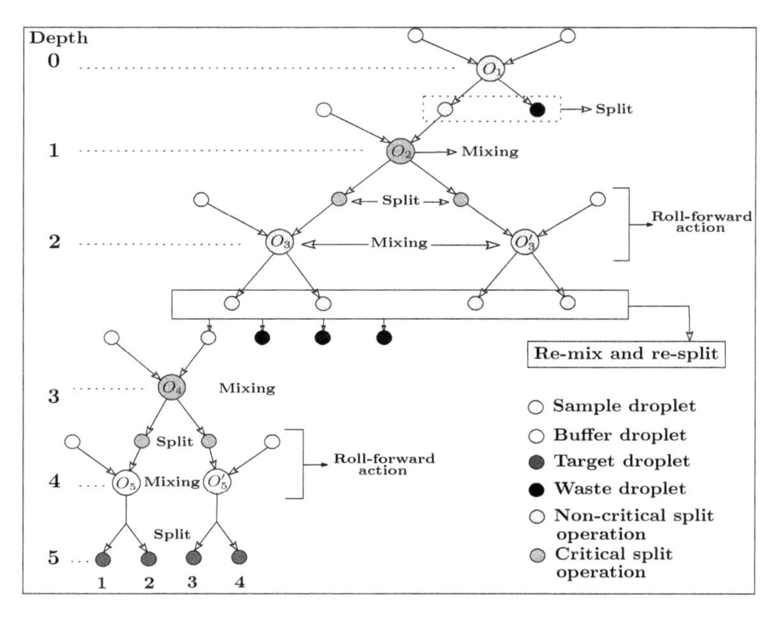

Figure 6.10: Error-tolerant dilution tree for generating a target-*CF* for accuracy level 5 using EOSP (with permission from IEEE [124])

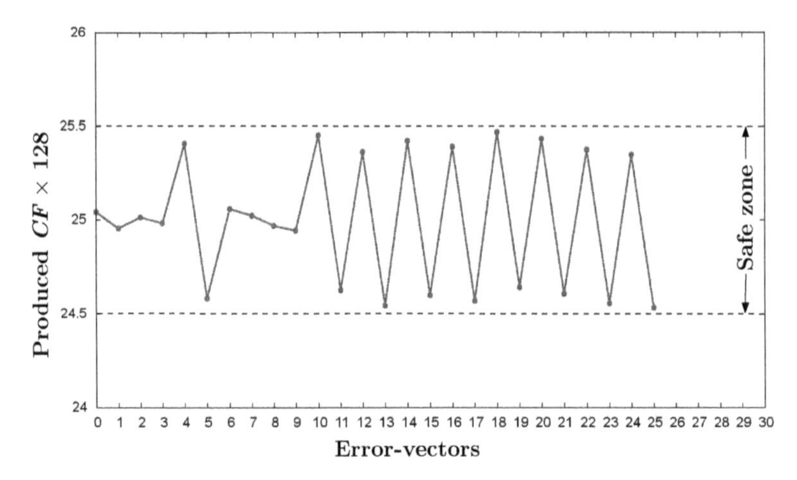

Figure 6.11: *CF*-values generated by the error-tolerant graph in the presence of multiple split-errors (with permission from IEEE [124])

should be shifted away from the split-location and the other one is moved towards the cell where the split-operation was performed. As a result, the residue will be merged with latter droplet, and hence, finally, only two daughter-droplets will remain. Any volumetric imbalance, if present, can be then handled using EOSP.

6.6 EXPERIMENTAL RESULTS

EOSP has been implemented in Python and on a computing platform with 2 GHz Intel Core i5 processor, 8 GB memory, and 64-bit Ubuntu 14.04 operating system. The results are compared with a baseline approach and conventional rollback error-recovery approaches [105, 192]. In the baseline approach, a complete dilution tree of depth n is generated and all the droplets produced at leaf nodes are mixed together to prepare a solution with the desired target-*CF* (Fig. 6.7). This process produces the desired target-*CF* accurately (*CF* error = 0) regardless of the occurrence of any volumetric split-error in the mixing tree (Section 2). For the rollback procedure, we consider the dilution tree [163] for a given target-*CF*, assume that two check-points are assigned (one near the middle and another at the end) for sensing the volume of split-droplets. Thus, mixing-path of each target-*CF* is divided into two segments. If an on-chip sensor detects an error at the end of a segment, then operations are re-executed starting from the initial position of that segment. Note that every error-recovery attempt increases reactant-cost and assay-completion time significantly. We performed experiments for some representative target-*CF*s and observed the cost and time overhead. For several error-recovery attempts in each segment, we report the results in Figs. 6.12 and 6.13. It can be observed from Fig. 6.12 that the

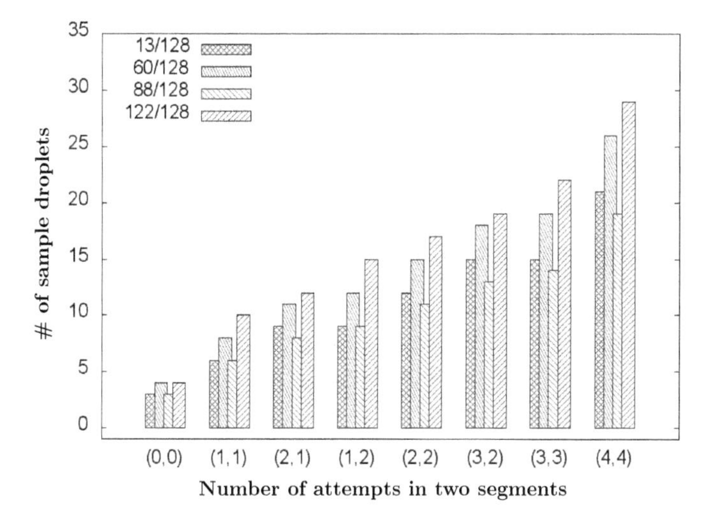

Figure 6.12: Consumption of sample droplets for preparing different target-*CFs* with accuracy level 7 by the rollback procedure (with permission from IEEE [124])

consumption of sample-droplets increases considerably when the number of recovery attempts/segment increases. Furthermore, sensing for volumetric imbalance at

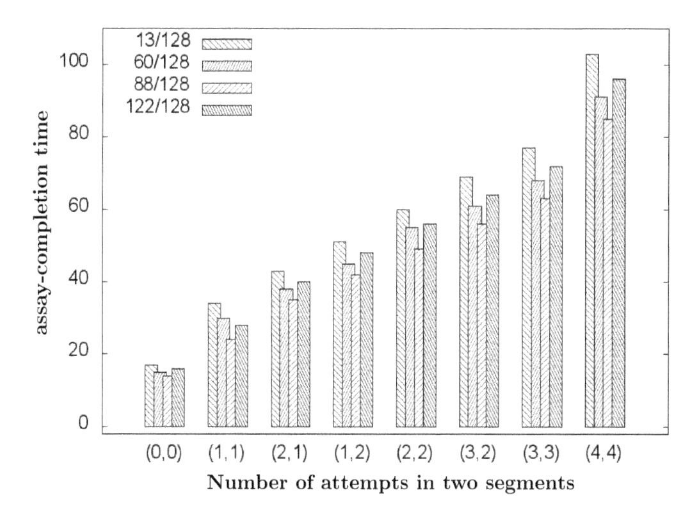

Figure 6.13: Generation time for different target-*CF*s with accuracy level $n = 7$ by the rollback procedure (with permission from IEEE [124])

the checkpoints requires additional time. For an illustration, we assume that two 1X volume droplets are mixed in 3 units of time and on-chip sensing including transportation requires 5 units of time. From Fig. 6.13, it can be seen that the assay-completion time for rollback approaches increases when the number of recovery-attempts per segment increases.

We performed various experiments on different real-life test-cases as well as synthetic test-cases. Some real-life test cases include: 95% ($CF = \frac{122}{128}$) *ethanol* is required in the glucose-tolerance test for mice and total *RNA* extraction from worms; a sample of 10% ($CF = \frac{13}{128}$) fetal bovine serum (*FBS*) is required for *in vitro* culture of human peripheral blood mononuclear cells (*PBMCs*) [17]; Bradford protein assay requires (10%, 20%, 30%, 40%, 50%) dilutions of a sample [17]. We simulated the baseline approach, the rollback-based error-recovery, and EOSP on these

Table 6.4
Performance evaluation for different methods.

Real-life test-cases							
CF	Baseline			EOSP [124]			$\lvert CF-\overline{CF}\rvert < \frac{0.5}{128}$?
	$\overline{CF}$	$\lvert CF-\overline{CF}\rvert$	n_d	$\overline{CF}$	$\lvert CF-\overline{CF}\rvert$	n_d	
$\frac{13}{128}$	$\frac{13}{128}$	0	128	$\frac{12.87}{128}$	$\frac{0.13}{128}$	16	yes
$\frac{27}{128}$	$\frac{27}{128}$	0	128	$\frac{26.79}{128}$	$\frac{0.21}{128}$	16	yes
$\frac{38}{128}$	$\frac{38}{128}$	0	64	$\frac{37.89}{128}$	$\frac{0.11}{128}$	4	yes
$\frac{51}{128}$	$\frac{51}{128}$	0	128	$\frac{50.85}{128}$	$\frac{0.15}{128}$	8	yes
$\frac{122}{128}$	$\frac{122}{128}$	0	64	$\frac{122.35}{128}$	$\frac{0.35}{128}$	8	yes
Synthetic test-cases							
CF	Baseline			EOSP [124]			$\lvert CF-\overline{CF}\rvert < \frac{0.5}{128}$?
	$\overline{CF}$	$\lvert CF-\overline{CF}\rvert$	n_d	$\overline{CF}$	$\lvert CF-\overline{CF}\rvert$	n_d	
$\frac{43}{128}$	$\frac{43}{128}$	0	128	$\frac{42.63}{128}$	$\frac{0.37}{128}$	4	yes
$\frac{65}{128}$	$\frac{65}{128}$	0	128	$\frac{64.90}{128}$	$\frac{0.10}{128}$	16	yes
$\frac{77}{128}$	$\frac{77}{128}$	0	128	$\frac{76.83}{128}$	$\frac{0.17}{128}$	8	yes
$\frac{90}{128}$	$\frac{90}{128}$	0	64	$\frac{89.98}{128}$	$\frac{0.12}{128}$	4	yes
$\frac{105}{128}$	$\frac{105}{128}$	0	128	$\frac{104.61}{128}$	$\frac{0.39}{128}$	16	yes

$\overline{CF}$: generated CF; n_d: the number of generated droplets.

test-cases and report the results in Table 6.4 and 6.5. During experiments, we randomly injected a number of volumetric split-errors on different mix-split steps of the mixing graph and noted various parameters, e.g., usage of reactants (sample and buffer), waste droplets, count of (1:1) mix/split operations. The error-rate (i.e., the number of error-occurrences on the mix-split steps) is an important parameter for the rollback approach; this determines how many times the rollback mechanism is to be invoked, and consequently, the cost for implementing the corrective action. For the rollback approach, we assumed an optimistic situation, where split-errors, once

Table 6.5
Performance for different methods.

Real-life test-cases

CF	Rollback				Baseline				EOSP [124]			
	n_m	n_s	n_b	n_w	n_m	n_s	n_b	n_w	n_m	n_s	n_b	n_w
$\frac{13}{128}$	(14,25,32,43)*	(06,12,15,21)	(10,17,22,29)	(04,27,35,48)	127	13	115	0	18	3	16	3
$\frac{27}{128}$	(14,27,29,39)	(06,15,18,26)	(08,14,16,24)	(14,27,32,48)	127	27	101	0	18	5	14	3
$\frac{38}{128}$	(12,19,27,36)	(06,09,14,19)	(08,12,18,24)	(12,19,30,41)	63	19	45	0	8	3	6	5
$\frac{51}{128}$	(14,21,32,43)	(08,12,18,24)	(08,12,19,26)	(14,22,35,48)	127	51	77	0	11	3	9	4
$\frac{122}{128}$	(12,21,27,36)	(10,17,22,29)	(04,08,10,14)	(12,23,30,41)	63	61	3	0	10	10	2	4

Synthetic test-cases

CF	Rollback				Baseline				EOSP [124]			
	n_m	n_s	n_b	n_w	n_m	n_s	n_b	n_w	n_m	n_s	n_b	n_w
$\frac{43}{128}$	(04,25,32,43)*	(08,15,19,26)	(08,14,18,24)	(14,27,35,48)	127	43	85	0	9	4	6	6
$\frac{65}{128}$	(14,25,32,43)	(04,06,09,12)	(12,22,28,38)	(14,26,35,48)	127	63	65	0	18	9	10	3
$\frac{77}{128}$	(14,25,32,43)	(08,12,19,26)	(08,14,18,24)	(14,27,35,48)	127	77	51	0	11	7	5	4
$\frac{90}{128}$	(12,21,27,36)	(08,14,18,24)	(06,11,14,19)	(12,23,30,41)	63	13	51	0	8	4	5	5
$\frac{105}{128}$	(14,25,32,43)	(08,14,18,24)	(08,15,19,26)	(14,27,35,48)	127	105	23	0	19	14	6	4

n_m: the number of (1:1) mixing; n_w: the number of waste droplets; $n_s(n_b)$: the number of sample (buffer) droplets; * for error-rate = (1, 2, 3, 4).

being sensed, is corrected within at most two (three) attempts in each segment, i.e., having error-rate 2 (3).

We consider accuracy level $(n) = 7$, and report the generated *CF*-values, the number of target-droplets (load), *CF*-error, and the error-limit for each target-*CF* in Table 6.4. We generate each target-*CF* multiple times (minimum 20 times and maximum = 50 times) by running EOSP and report the maximum *CF*-error in Table 6.4. Note that the target-droplets generated by the existing rollback approach may have some small amount of *CF*-error. This may be caused by some split-induced volumetric imbalance, which is not detectable by the sensor. Hence, it is not possible to compute the generated *CF*-values accurately for rollback approaches. We report these results for the baseline and the EOSP method only. The latter reduces the number of target-droplets significantly compared to baseline, yet efficiently bounds the target-*CF*-error within the allowable error-limit, in a deterministic fashion, and without the need for any sensing.

The required number of sample (buffer) droplets, waste droplets, and number of (1:1) mix-split operations for each of the target-*CF*s are reported in Table 6.5. For the rollback approach, results are shown for error-rate 1, 2, 3 and 4. It can be

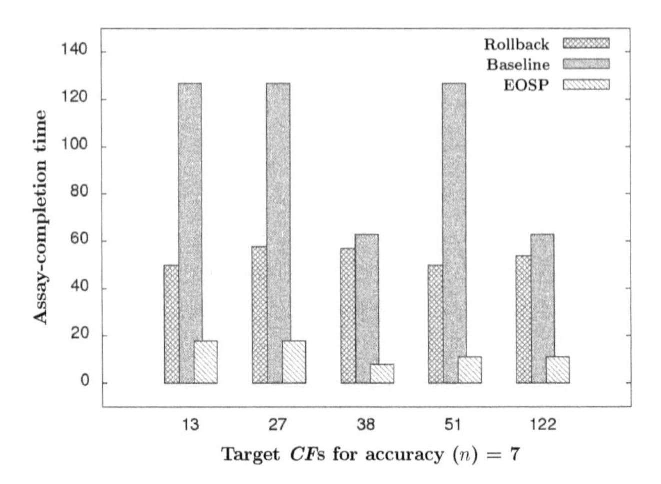

Figure 6.14: Total time required to generate the desired target-*CF*s of accuracy level 7 by various methods (with permission from IEEE [124])

observed that EOSP minimizes all these parameters noticeably compared to the baseline and rollback methods. Also, the rollback-based methods generate more intermediate waste droplets than EOSP. For some representative test-cases, assay-completion times needed by rollback, baseline, and EOSP are depicted in Fig. 6.14.

We perform experiments for finding the minimum number of target droplets that are to be generated (output load) to achieve the error-tolerance for each target-*CF* having accuracy-level = 7. While generating the target-*CF*s, we inject a fixed percentage of volumetric split-error on each mix-split step of the mixing/reaction path and record the minimum load needed by each target-*CF*. We observe that the minimum number of target-droplets needed to cancel 3% of the split-error is 8, whereas it increases to 16 when 5% and 7% split-error are to be tolerated, respectively. Interestingly, a large number of the target-*CF*s (80%) require only 8 target-droplets for canceling up to 7% volumetric split-error. We also note that mixing or splitting is not required after generating the target droplets (only dispensing is required from the target-reservoir). However, larger volume droplets (4X, 8X, 16X) need to be mixed when one or more critical error occur consecutively on the mixing path before reaching the final step. We have also performed experiments to study their frequencies for all target-*CF*s with accuracy level = 7 and the results are shown in Fig. 6.15 for 3%, 5%, and 7% multiple volumetric split-errors.

Note that the area of the reservoir (for temporarily holding/storing droplets) is determined by the maximum number of intermediate droplets generated by consecutive critical segments. We run certain experiments and observe that at most three consecutive critical steps appear for accuracy level ≤ 7. Hence, one 4X-, 8X-, and 16X-size, i.e., a total of three reservoirs are sufficient for temporarily storing all intermediate droplets on the chip. After mixing operations, droplets can be placed into the reservoir and a single droplet can be easily egressed from the reservoir with just

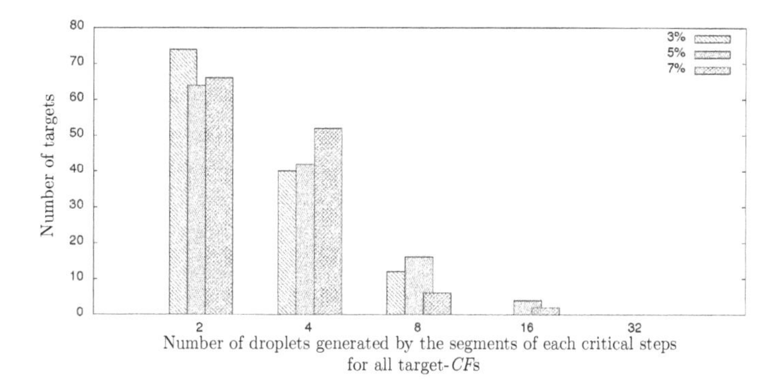

Figure 6.15: Number of target-*CF*s that require reservoirs of size 4X, 8X, and 16X for storing intermediate droplets, when accuracy level = 7 and ε is varied (with permission from IEEE [124])

one dispensing operation. This type of dispensing operation can be performed using some prior techniques for DMFBs [11, 27, 33, 75, 83, 131]. Such a dispensing operation minimizes the cumulative split-error that may otherwise crop in when a 1X volume-droplet is generated by sequentially splitting into two parts, in each splitting step, a larger-volume droplet such as 16X, 8X, or 4X.

We compare the performance of EOSP with the roll-forward approach [122] and report results in Table 6.6 under different error-rates. During experiments, we ran-

Table 6.6
Comparison of EOSP [124] with the roll-forward method [122].

Target-*CF*	Methods	$E_r = 0$*				$E_r = 1(2)$				$E_r = 3(4)$			
		n_s	n_b	n_w	e_t	n_s	n_b	n_w	e_t	n_s	n_b	n_w	e_t
$\frac{27}{128}$	[122]	4	4	6	36	5(5)	6(10)	7(7)	55(77)	5(5)	12(14)	5(3)	83(89)
	[124]	5	14	3	54	5(5)	14(14)	3(3)	54(54)	5(5)	14(14)	3(3)	54(54)
$\frac{47}{128}$	[122]	5	3	6	31	6(6)	5(5)	7(7)	45(45)	6(6)	7(7)	5(5)	51(51)
	[124]	6	7	9	36	6(6)	7(7)	9(9)	36(36)	6(6)	7(7)	9(9)	36(36)
$\frac{83}{128}$	[122]	4	4	6	31	6(6)	5(5)	7(7)	45(45)	8(8)	5(5)	5(5)	51(51)
	[124]	7	3	6	27	7(7)	3(3)	6(6)	27(27)	7(7)	3(3)	6(6)	27(27)
$\frac{100}{128}$	[122]	3	3	4	30	5(9)	4(4)	5(5)	49(71)	11(13)	4(4)	3(1)	77(83)
	[124]	13	4	1	48	13(13)	4(4)	1(1)	48(48)	13(13)	4(4)	1(1)	48(48)
$\frac{113}{128}$	[122]	4	4	6	36	7(11)	4(4)	7(7)	55(77)	13(15)	4(4)	5(3)	83(89)
	[124]	15	5	4	57	15(15)	5(5)	4(4)	57(57)	15(15)	5(5)	4(4)	57(57)

E_r: Error rates, $n_s(n_b)$: the number of consumed sample (buffer) droplets, n_w: the number of waste droplets, e_t: (mixing time + sensing time) in sec, * ideal scenario.

domly inject a number of volumetric split-errors (7%) in the mix-split steps of the dilution tree for five representative target-*CF*s and report the number of sample (buffer) droplets, waste droplets, and the overall assay-completion time (mixing time + sensing time) in Table 6.6. We assume that a (1:1) mixing requires 3 sec and a detection operation requires 5 sec (including the time needed for droplet-transportation from the mixer to the sensor-location). It can be observed from Table 6.6 that the roll-forward method [122] consumes a smaller amount of reactant (sample and buffer) compared to EOSP in an ideal-scenario (in the absence of the splitting-error). However, the consumption gradually increases with increasing error-rate, and both methods consume an almost equal amount of reactants when error-rate becomes 4. In the presence of multiple split-errors, the sensing time will predominate and the total assay-completion time may increase for the roll-forward method and EOSP clearly outperforms it. It is reported in the literature that long sample-preparation time may cause a delay in diagnosis and treatment [20,41,82,113,144,196].

Table 6.7
Variation of error-tolerance ($\frac{x}{2^{n+1}}$) limit for synthetic test-cases.

Target-*CF*	$x=1$				$x=2$				$x=3$				$x=4$			
	n_s	n_b	n_m	n_d	n_s	n_b	n_m	n_d	n_s	n_b	n_m	n_d	n_s	n_b	n_m	n_d
$\frac{47}{128}$	6	7	13	4	5	5	9	4	5	4	8	4	5	4	8	4
$\frac{65}{128}$	9	10	18	16	5	7	11	8	3	7	9	4	3	6	8	4
$\frac{99}{128}$	14	5	18	16	8	4	11	8	7	5	11	2	5	4	8	2
$\frac{105}{128}$	14	6	19	16	8	4	11	8	6	4	9	2	5	4	8	2

$n_s(n_b)$: the number of sample (buffer) droplets, n_m: the number of (1:1) mix-split steps, n_d: the number of generated target-droplets.

So far we have assumed a strict condition on error-tolerance, i.e., the *CF*-error in the target-droplet must be less than $\frac{0.5}{2^n} \times x$, where $x = 1$. However, it (error-tolerance) can be relaxed in some scenarios (e.g., in real-life biochemical experiments [89]), i.e., x may become greater than 1. Accordingly, we vary x from 1 to 4 in our experiments and report the reactant-cost, generated target-droplets, and the number of (1:1) mix-split steps in Table 6.7. We observe that the sample-preparation cost with error obliviousness reduces noticeably when the tolerance is increased. Further, it minimizes the number of (1:1) mix-split operations, which in turn, reduces sample-preparation time.

6.7 CONCLUSIONS

In this chapter, we have introduced a robust sample-preparation method (EOSP) based on the concept of error obliviousness for digital microfluidic lab-on-chips. Only pre-computed actuation sequences are required to govern the execution of the dilution assay regardless of the presence of any volumetric split-error, single or multiple. The robustness of EOSP has been validated through rigorous simulation experiments and case-studies. EOSP additionally allows for a reduction in reactant cost and mix-split operations in the presence of split-errors compared to other approaches. Also, it requires minimal resources and can be implemented on a DMFB without cyber-physical extensions such as sensors and feedback mechanisms. As a future research work, one can study how error-oblivious mixture preparation can be automated on a biochip.

7 Robust Multi-Target Sample Preparation On-Demand with DMFBs

Most of the bio-protocols (e.g., those required for clinical diagnostics and point-of-care (*PoC*) applications) consist of the following four sequential steps: sample collection, sample preparation, analytical processing, and detection (sensing). Over the years, digital microfluidic technology has been widely used to implement automatic sample preparation [100, 122, 131, 135, 163], analytical processing [26], and detection mechanism [104]. Thus, sample preparation has become an essential step in almost all bio-protocols. Various optimization goals during sample preparation include the minimization of the reagent-cost, waste-droplets, and mix-split count (i.e., sample-preparation time) [15]. Dilution is the special case of mixing two different types of bio-fluids (sample and buffer) with a certain volumetric ratio such that a desired concentration factor (*CF*) of the sample is prepared.

Multiple concentration factors of a sample may become necessary in many real-life applications. For example, (i) samples and reagents are required in multiple concentration (or dilution) factors, satisfying certain "gradient" patterns in bacterial susceptibility tests for drug design [141]. The purpose is to determine the minimum amount of an antibiotic that will inhibit the visible growth of a microorganism/bacteria-isolate (also known as Minimum Inhibitory Concentration (MIC)). The drug with the lowest concentration (highest dilution) that will prevent the growth of an organism/bacteria satisfies the condition of MIC [28, 160]. Multiple target concentrations of a reagent are also needed in protein crystallization and DNA analysis [72, 149, 184]. Furthermore, multiple droplets of different concentrations are required in order to reduce inflammatory mediators such as cytokines and other circulating inhibitors or binding proteins, with high sensitivity [90]. Samples with different concentrations are also needed while executing several assays on a multiplexed biochip in a concurrent fashion. Additionally, sample-preparation time needs to be reduced since long sample-processing time may cause a delay in diagnosis, which may turn out to be catastrophic for certain real-life biochemical assays. Also, in order to prevent degradation or contamination, collected blood samples [196] and gene expression profiling samples (RNA isolation) need to be processed promptly [41] without any delay. Similarly, bioassays need to be completed rapidly while performing toxicity analysis, forensics [20, 144], or molecular detection of antibiotic resistant pathogens in human urine [167]. Therefore, while performing sensitive tests, all necessary samples must be prepared quickly in order to prevent any delay in processing or diagnosis.

DOI: 10.1201/9781003219651-7

In the case of multiple-target sample preparation (dilution), there is another important issue concerning the optimization of sample-preparation time and reagent-cost. During multi-target sample preparation on a DMFB platform, the waste-droplets that are produced at different mix-split steps are often shared to reduce assay cost and time [57, 100, 113, 163]. In order to implement such procedures, the set of target-*CF*s is assumed to be known in advance which may not always become possible. For example, in many practical settings, the demand for target-*CF*s may arrive sequentially on-the-fly, i.e., the information about the complete set of target-*CF*s to be produced, is not known *a priori*. Unfortunately, all previous methods fail to solve such optimization problems efficiently for these instances.

The correctness or the accuracy of target-*CF*s depends on the outcome of split-operations executed during sample preparation. The target solutions may not become acceptable when the error in *CF* exceeds the allowable threshold limit. Many life-critical applications, such as clinical diagnostics, DNA analysis, or drug design, require high precision where the outcome of an assay cannot be compromised [124]. Conventional error-recovery approaches [105–107, 192] cannot handle the production of multiple target-*CF*s reliably and efficiently. The identification of critical (non-critical) nodes and the construction of error-tolerant graphs corresponding to the multi-target dilution problem become far more complex due to the presence of fork nodes (where waste-droplets are shared) in the dilution-tree. Although EOSP (presented in Chapter 6) can be extended in principle to handle the multi-target problem, a rigorous analysis is required to redefine the notion of criticality of spilt-errors and to assess their impact on the target-*CF*s.

In this chapter, we present a cost-effective solution to Robust Multi-Target Sample Preparation On Demand (MTD) that requires only one (or two) mix-split step(s) [121]. In order to service dynamically-arriving requests for multiple *CFs* quickly, we prepare dilutions of the sample with a few *CFs* in advance (called source-*CF*s), and fill on-chip reservoirs with these fluids. For minimizing the number of such preprocessed *CFs*, we present an ILP-based method, an approximation algorithm, and a heuristic algorithm. These methods also allow the users to trade-off the number of on-chip reservoirs against service time for various applications. Simulation results for several target sets demonstrate the superiority of MTD over prior art in terms of the number of mix-split steps, waste droplets, and reactant usage when the on-chip reservoirs are pre-loaded with source-*CF*s using a customized droplet-streaming engine. In addition, we also present an alternative method for producing multi-target-*CF*s, reliably, even when mix-split operations are performed in an error-prone environment.

The remainder of the chapter is organized as follows. The motivation behind MTD is presented in 7.1. Section 7.2 briefly describes the basics of sample preparation. Section 7.3 summarizes prior art on sample preparation/dilution with DMFB. Section 7.4 introduces the problem of on-demand dilution, presents formulation of the problem and lists our contributions. Solution approaches to rapid and on-demand sample preparation are presented in Section 7.5. Section 7.6 discusses another variant of the problem. The impact of the number of reservoirs on the performance of MTD is addressed in Section 7.7. A low-cost streaming engine for filling the on-chip reservoirs is presented in Section 7.8 Experimental results are reported in Section 7.9.

Error-free multi-target dilution, on-demand, is discussed in Section 7.10. Finally, conclusions and discussion on open problems are presented in Section 7.11.

7.1 MOTIVATION

Poddar *et al.* [121] described a new approach to solve the above-mentioned "on-demand" sample preparation problem for multiple target-*CF*s. As by-product, the method can also be used for rapid delivery of the desired target droplets with reduced cost. We list below a few examples of biochemical protocols where on-demand or rapid sample preparation is mandated.

1) Genomics using single-cell analysis (i.e., extracting information from single cells for understanding the cell population) has gained significant momentum in recent times in tandem with the rapid advancement of microfluidic technologies, which are affordable and can provide results accurately and efficiently [54]. Microfluidic biochips can be utilized to simultaneously explore thousands of heterogeneous cells. Thus, the link between gene expression and cell types can be easily discovered, thereby providing useful insights into diseases such as cancer [137]. On-demand sample preparation is required during single-cell analysis, which is carried out on the constituent cell. In the first-step, a bio-protocol is selected in real-time depending on the cell type [54]. For instance, for certain cell type A, a laboratory investigator may want to examine the gene-expression level for a specific genomic region. At the same time, for a different cell type B, another specific location of the genome may need to be explored to identify protein interactions (i.e., causative proteins) that lead to an unexpected heterochromatic state in the corresponding region (loci). Although a quantitative-PCR analysis may become sufficient for the specific genome analysis of cell type A, chromatin immune-precipitation (ChIP) followed by quantitative-PCR may become necessary for cell-type B in order to reveal the DNA-strains. So, bioassays need to be selected in real-time depending on the cell-type (when thousands of demand cases arrive on-the-fly [148]). Furthermore, fast completion of ChIP is necessary to enable the downstream analysis, e.g., on-demand polymerase chain reaction [148]. Several reagents (e.g., formaldehyde, antibodies) [2,71,80] with different concentration factors are also required for performing ChIP [2, 71, 80]. Therefore, depending on the cell type, requests for supplying multiple target-*CF*s may be made in real-time [148]. Another highly effective technique for accurately monitoring and identifying bacteria and viruses [148] is on-demand PCR-testing. Rapid and accurate diagnosis (following the collection of a biochemical sample from a patient) can shorten the duration of hospital stay for patients an average of two days, reduce the mortality rate from 9.6% to 7.9%, and result in annual cost savings [37]. It facilitates the initiation of personalized therapy at the bedside. However, in order to serve the purpose, primers with multiple *CF*s are required for the optimization of the real-time PCR (RT-qPCR) [110].

2) Many bio-protocols such as PCR and colorimetric assays are executed following a single pathway. However, in more complex assays, e.g., quantitative gene expression, the actual sequence of fluidic operations remains unknown until intermediate reaction results are available [69]. In other words, the underlying task-graph of such bioprotocols mimics an "if-then-else" condition. DMFBs with integrated sensor facilitates are needed to check the outcome of intermediate operations at various checkpoints, and to execute a reaction-path based on the sensor feedback [69]. For example, some operations associated with ChIP need to be re-executed starting from cell lysis in the presence of nucleotides (NA) in the solution even after isolating it [67–69]. This type of bioprotocol requires various reagents (e.g., Triton X-100, SDS, NaCl, formaldehyde, antibodies) with different concentration factors, on-demand, which need to be prepared rapidly [18, 80, 81, 110]. Also, primers with different concentrations are required on-demand for DNA-amplification of multiple target sequences simultaneously or for re-executing the bioprotocol (in the case of insufficient amplification) [69]. Clearly, their requirement cannot be predicted in advance. Therefore, in such scenarios, rapid sample preparation with real-time demand turns out to be a key requirement.

3) On-demand sample preparation is also required in other applications such as relative potency (REP) estimation (widely used for comparing the potency of a wide variety of samples required in toxicology and drug design). *In-vitro* bioassays (powerful tool for environmental monitoring) are generally performed for REP estimation. The notation REP_x denotes the *CF* of the sample (drug) that is needed to elicit $x\%$ response in a test system or on organisms [172]. An experiment to conduct REP_{20-80} based on multiple point estimation requires multiple concentration factors of a dose. It measures the effect of a particular dose at different points. Based on the response, if required, it re-executes the same experiment with different dose-levels [172]. Clearly, this type of experiment requires different concentrations on-demand in real-time. Moreover, multiple-droplets with the same *CF* are also required for conducting such tests. Rapid on-chip sample preparation (multiple *CF*s of a sample) is also required in the study of molecular kinematics [146], which are needed for understanding protein (RNA) folding, enzymatic and chemical reactivity.

4) The MTD method [121] is also useful in producing multiple droplets with a given *CF* rapidly [133]. A stream of droplets with the desired target-*CF* may be required when (a) a bioassay is repeatedly executed for screening several patients in a point-of-care diagnostics center, (b) an assay is re-executed for verifying a test outcome, and (c) multiple droplets of the same target-*CF* are needed for performing several bio-analytical tests. A universal microfluidic platform capable of generating various concentrations of a sample was also designed for performing various experiments on-demand [76]. MTD produces very few waste droplets compared to other state-of-the-art dilution methods [57, 100] and hence it supports green chemistry [76].

7.2 BASICS OF SAMPLE PREPARATION

Dilution of a sample and mixing of reagents are the two key steps in any biochemical sample preparation. In mixture preparation, two or more reagents are mixed in a certain ratio, whereas in dilution process, a sample fluid is mixed with a buffer fluid only. In a diluted solution, the concentration factor (*CF*) is defined as the amount of raw sample present in a target droplet ($0 \leq CF \leq 1$) [163]. Note that *CF* of raw-sample (buffer) solution is assumed to be 1 (0). In a DMFB-platform, typically the (1:1) mixing model is adopted where a mix operation is performed between two unit-volume (1X-size) droplets followed by a balanced splitting into two 1X-size droplets; for example, if a unit-volume fluid droplet with $CF = C_1$ is mixed with another unit-volume fluid droplet with $CF = C_2$, then after the mix-split operation, the *CF* of each sister droplet becomes $\frac{C_1+C_2}{2}$, i.e., the mean of C_1 and C_2. In general, a solution with a desired target-*CF* can be produced by performing a sequence of (1:1) mix-split operations using sample/buffer droplets; the process is envisaged as a directed acyclic graph called *mix-split tree/graph* [163]. Note that the depth n (> 0) of the *mix-split graph* is determined by the user-defined error-tolerance limit τ ($0 \leq \tau < 1$) of *CF*s. In order to limit the error in target-*CF* by $\frac{1}{2^{n+1}}$, a sequence of at most n (1:1) mix-split operations is performed for achieving the target-*CF* in the *bit-scanning* method [163]. A value of $n = 8$ is found to be sufficient in all practical biochemical experiments that involves dilution [113]. Because of the (1:1) inherent mixing model supported by DMFB platform, each *CF* is required to be approximated as n-bit binary fractional number $\frac{x}{2^n}$, depending on τ; where $x \in \mathbb{N}$, $0 \leq x \leq 2^n$, and $n \in \mathbb{N}$ [15]. Note that for a given τ, the user needs to choose an n, so that the error in target-*CF* is upper-bounded by $\frac{1}{2^{n+1}}$. Since the number of mix-split operations influences sample-preparation time, a method with fewer mix-split operations will be preferable for rapid and on-demand sample preparation, for a given n.

7.3 LITERATURE REVIEW

A sample-preparation algorithm (*MinMix*) based on binary bit-scanning was proposed by Thies *et al.* [163] for mixing two or more sample/reagent fluids with a given ratio using the (1:1) mixing model. For the special case of dilution, *twoWayMix* [163] can generate a minimum-length mix-split sequence needed for producing a sample with a desired *CF* using sample and buffer droplets. Other dilution algorithms, namely Dilution/Mixing with Reduced Wastage (*DMRW* [131]) and Improved Dilution/Mixing Algorithm *IDMA* [135]) were proposed by Roy *et al.*, which aim to reduce the number of waste droplets. Subsequently, REactant-MInimization dilution Algorithms (*REMIA* [60]) and Graph-based Optimal Reactant-Minimization Algorithm (*GORMA* [25]) were reported for reducing the cost of reactants (sample/buffer).

Note that a sample-preparation algorithm for constructing a mix-split tree can be used to produce the sequence for diluting any fluid, in general. Thus, from the viewpoint of computer-aided design (CAD), we use the term reagent or reactant with a broader connotation that includes all fluids, over and above traditional biochemical samples and buffers.

For generating multiple target dilutions, several approaches have been reported in the literature. Bhattacharjee *et al.* [12] proposed a Pruning-Based Dilution Algorithm (*PBDA*) for generating multiple target-*CF*s based on the sharing of dilution sub-trees for minimizing the number of mix-split steps and waste droplets. The multi-target Reagent Saving Mixing (*RSM*) [57] algorithm not only minimizes the number of valuable reactant-droplets but also reduces waste production and the number of mixing operations. Huang *et al.* [100] developed a reactant-reducing multi-*CF* dilution algorithm, namely WAste-Recycling Algorithm (*WARA* [100]), in which the individual mixing trees generated by *REMIA* [60], are combined to provide efficient droplet-sharing and waste-recycling. Mitra *et al.* [113] proposed an algorithm called Multiple-Target Concentration (*MTC*) under storage-constraints. Recently, a network-flow based optimal sample-preparation algorithm has been developed by Dinh *et al.* [31] for producing multiple target-*CF*s (each having a non-zero demand of droplets).

In Table 7.1, we have summarized the key features of different dilution algorithms for DMFB-platforms based on their target-profile and optimization objectives (reduction of mix-split steps/waste/reactant-cost). Additionally, we have highlighted their attributes in regard to the capability of on-demand supply and service time. Note that sample-preparation (service) time increases with decreasing error-tolerance allowed in *CF*; also for muti-target-*CF* generation, sample-preparation time will increase further when target-*CF*s are not known *a-priori*.

Table 7.1
Scope of various dilution algorithms

Dilution algorithm	Produces multiple target-CFs?	Aims to reduce #mixing steps?	Aims to reduce #waste-droplets?	Aims to reduce #reactant-droplets?	On-demand?	Rapid service?	Error oblivi-ous?
twoWayMix [163]	no	yes	no	no	no	no	no
DMRW [131]	no	no	yes	no	no	no	no
IDMA [135]	no	yes	yes	no	no	no	no
REMIA [60]	no	yes	no	yes	no	no	no
GORMA [25]	no	yes	no	yes	no	no	no
PBDA [12]	yes	yes	yes	no	no	no	no
RSM [57]	yes	yes	yes	no	no	no	no
WARA [100]	yes	yes	yes	yes	no	no	no
MTC [113]	yes	yes	yes	yes	no	no	no
Dinh *et al.* [32]	yes	yes	yes	yes	no	no	no
MTD [121]	yes	yes	yes	yes	yes	yes	yes

7.4 ON-DEMAND MULTI-TARGET DILUTION-PROBLEM (MTD)

As discussed in the previous section, there exist a number of sample-preparation algorithms that are capable of producing single/multi-*CF* targets considering several optimization objectives such as minimization of the number of mix-split steps, waste-production, and reactant-cost. We demonstrate below, using experimental data, how these optimization goals are severely degraded when existing algorithms [57,100] for producing multi-*CF* dilutions are invoked in the case when the target-set is not fully known in advance. We consider a multi-*CF* target set of size 100 with $n = 8$ and partition it into differently-sized disjoint subsets (window size) having cardinality of 1, 5, 10, 20, 50, 100. We assume that the requests for producing these subsets of target-*CF*s arrive sequentially (on-demand). The performance of *WARA-sharing-only*[1] [100] and *RSM* [57] with respect to varying window-sizes is shown in Table 7.2. The top-most (bottom-most) row in each of the two groups corresponds to the window-size 100 (1), representing the instance where all elements (only one element) of the target set are (is) known to the laboratory-service provider. Note that as the window-size reduces, all parameters (#mix-split steps, #waste-droplets, #sample/buffer droplets) increase significantly. Hence, both sample-preparation cost and time are adversely affected when target-*CF*s are requested on-demand. To explain the context, let us present a motivating example.

Table 7.2
Performance of multi-*CF* dilution algorithms based on *a priori* knowledge of window size.

Dilution Algorithm	window-size × #windows	#(1:1) mix-split steps	#sample droplets	#buffer droplets	#waste droplets
WARA-sharing-only [100]	100 × 1	288	109	112	121
	50 × 2	313	114	124	138
	20 × 5	387	133	167	200
	10 × 10	450	147	205	252
	5 × 20	521	168	250	318
	1 × 100	763	233	463	596
RSM [57]	100 × 1	211	109	106	115
	50 × 2	257	136	126	162
	20 × 5	344	180	174	254
	10 × 10	409	214	208	322
	5 × 20	501	265	258	423
	1 × 100	695	400	395	695

Example 11 Let $n = 4$, i.e., each *CF*-value is approximated as 4-bit fractional binary number, where the boundary-*CFs* 0 and 1 correspond to the concentration of buffer and raw sample, respectively. The other *CF*-values belong to the set $T_4 = \{\frac{1}{2^4}, \frac{2}{2^4}, \frac{3}{2^4}, \ldots, \frac{2^4-1}{2^4}\}$. We want to partition the set T_4 into two subsets: T_{4A} and T_{4B} such that any $CF \in T_{4B}$ can be produced by applying *only one* (1:1)-mix operation on a pair of *CF*s belonging to T_{4A}. One may naively include in T_{4A}, all *CF*-values that have an even number in their numerator.

Fig. 7.1 shows such a partition of T_4 where any element of T_{4B} can be produced from at least one droplet-pair belonging to T_{4A} in one (1:1)-mixing step. Note that

[1] WARA has been implemented considering droplet-sharing only

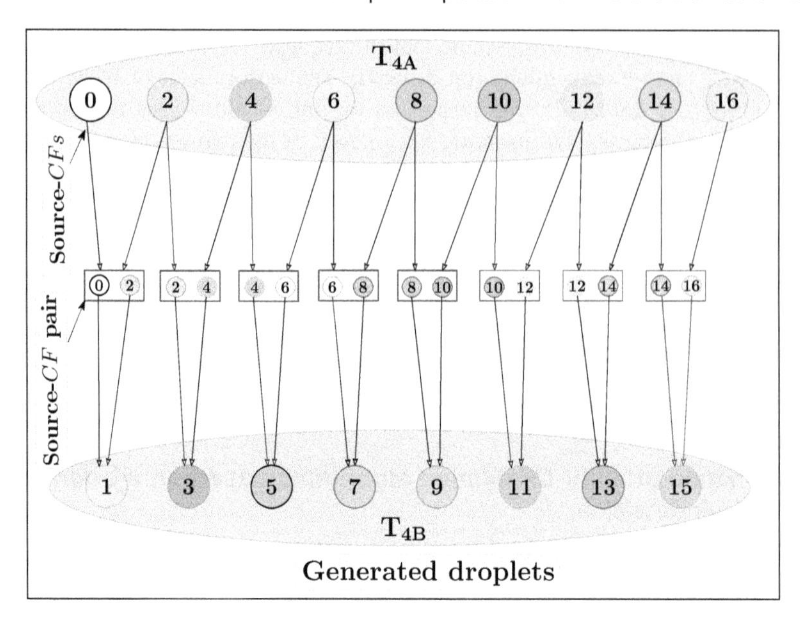

Figure 7.1: Naive solution of size 9 for $n = 4$ (with permission from IEEE [121])

for compactness of representation, only the numerators of *CF*-values are shown in the figure as nodes, and a pair of directed edges to a node in set T_{4B} indicates how the corresponding target-*CF* can be produced by mixing two source-droplets present in the set T_{4A}, in (1:1) ratio.

Interestingly, Fig. 7.2 shows another potential solution where T_{4A} comprises only six values of *CF* and the remaining *CF*-values in T_4 can be produced as before from it using only one mixing step. In general, the difficult question is how to determine a minimum-size subset of *CF*s such that any other *CF* belonging to the remaining set can be produced using only one (1:1) mix-operation. ■

We now formally introduce the problem of on-demand sample preparation and present a concise mathematical formulation. We assume that each *CF* is represented as an n-bit fractional binary number, where $n \in \mathbb{N}$ and $n \geq 1$. Hence, the concentration factor (*CF*) of the raw sample can be treated as 1 and the *CF* of buffer as 0. The remaining range of the *CF*s, for a given n, is denoted by $T_n = \{\frac{1}{2^n}, \frac{2}{2^n}, \frac{3}{2^n}, \cdots, \frac{2^n-1}{2^n}\}$.

7.4.1 PROBLEM DEFINITION

Given T_n, the problem is to determine the minimum cardinality subset $C_{min} \subset T_n$ such that for each $c \in T_n \setminus C_{min}$, there exists some $c_m, c_n \in C_{min}$ for which $c = \frac{c_m+c_n}{2}$. In other words, one should be able to prepare any sample with $CF \in T_n \setminus C_{min}$ from at least one pair of *CF*s belonging to C_{min} using only *one* (1:1)-mix operation. Therefore, only the samples with $CF \in C_{min}$ need to be stored ahead of time in order to produce any other *CF*s on-demand.

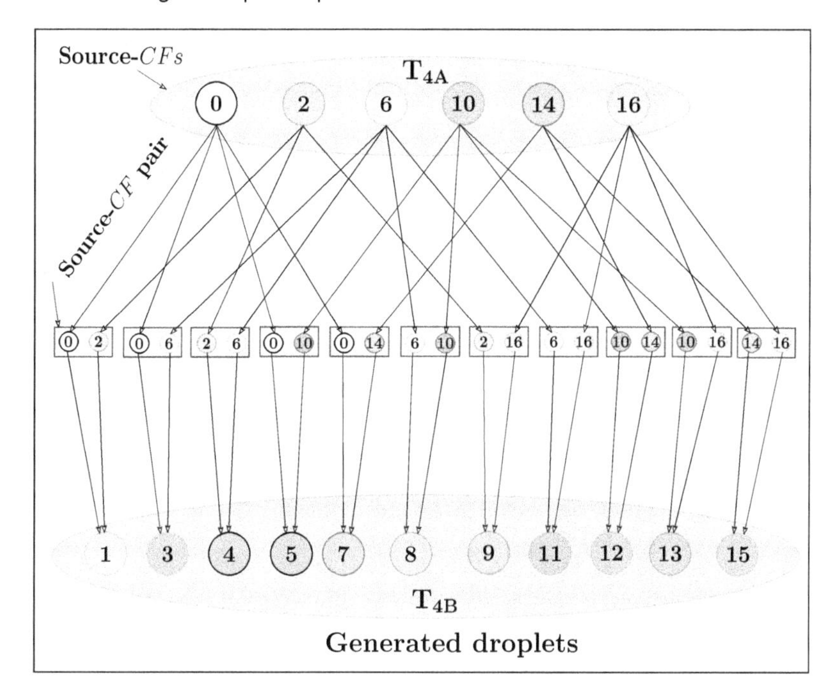

Figure 7.2: Improved solution of size 6 for $n = 4$ (with permission from IEEE [121])

Since a (1:1)-mix operation of two *CF*s yields a target whose *CF* is equal to the mean of the two source-*CF*s, the boundary-*CF*s, i.e., *CF* = 0 (smallest *CF* representing buffer) and *CF* = 1 (largest *CF* representing raw sample) must belong to C_{min}, as the smallest and the largest value can never appear as mean of two other elements of T_n.

7.4.2 MAIN RESULTS

Motivated by the necessity of handling multi-target sample preparation on-the-fly, MTD aims to produce a desired target-concentration using fewest on-chip reservoirs and only one or two mixing steps. The method includes the following:

1. A formulation based on integer linear programming (ILP) for determining the minimum number of source-*CF*s (= the number of on-chip reservoirs) that are to be stored for producing any other *CF* in T_n for a given n, in only one or two mix-split steps.
2. A constant-factor approximation algorithm for addressing the scalability issues confronted by the ILP formulation.
3. A fast heuristic algorithm for solving the same problem.
4. A reactant-saving algorithm for producing droplets with source-*CF*s, which are stored in on-chip reservoirs.

5. A study on the trade-off between the number of on-chip reservoirs and mix-split steps required for generating requested target concentrations on-the-fly.

7.5 RAPID PRODUCTION OF TARGET-CFs ON-THE-FLY

We first illustrate the underlying optimization problem with a simple example as follows.

Example 12 Let $n = 3$ and consider the problem of partitioning T_3 into two subsets C_{min} and $T_3 \setminus C_{min}$ such that any *CF* belonging to the latter subset can be produced via one (1:1)-mix operation on at least one pair of elements from C_{min}. Table 7.3 shows all possible ways of generating a given concentration factor via one (1:1)-mix operation. If we choose $C_{min} = \{\frac{0}{2^3}, \frac{2}{2^3}, \frac{6}{2^3}, \frac{8}{2^3}\}$, then each *CF* in $T_3 \setminus C_{min}$ can be generated by combining two concentrations from C_{min}. Note that the underscored entries in Table 7.3 indicate possible *CF*-pairs belonging to C_{min} that can be used to generate other *CF*s in $T_3 \setminus C_{min}$. ■

Table 7.3
Possible source-pairs of *CF*s for a target-*CF* in T$_3$.

Target	Possible pairs of *CF*-values
$\frac{1}{2^3}$	$\{\underline{(\frac{0}{2^3}, \frac{2}{2^3})}, (\frac{1}{2^3}, \frac{1}{2^3})\}$
$\frac{2}{2^3}$	$\{(\frac{0}{2^3}, \frac{4}{2^3}), (\frac{1}{2^3}, \frac{3}{2^3}), \underline{(\frac{2}{2^3}, \frac{2}{2^3})}\}$
$\frac{3}{2^3}$	$\{\underline{(\frac{0}{2^3}, \frac{6}{2^3})}, (\frac{1}{2^3}, \frac{5}{2^3}), (\frac{2}{2^3}, \frac{4}{2^3}), (\frac{3}{2^3}, \frac{3}{2^3})\}$
$\frac{4}{2^3}$	$\{\underline{(\frac{0}{2^3}, \frac{8}{2^3})}, (\frac{1}{2^3}, \frac{7}{2^3}), \underline{(\frac{2}{2^3}, \frac{6}{2^3})}, (\frac{3}{2^3}, \frac{5}{2^3}), (\frac{4}{2^3}, \frac{4}{2^3})\}$
$\frac{5}{2^3}$	$\{\underline{(\frac{2}{2^3}, \frac{8}{2^3})}, (\frac{3}{2^3}, \frac{7}{2^3}), (\frac{4}{2^3}, \frac{6}{2^3}), (\frac{5}{2^3}, \frac{5}{2^3})\}$
$\frac{6}{2^3}$	$\{(\frac{4}{2^3}, \frac{8}{2^3}), (\frac{5}{2^3}, \frac{7}{2^3}), \underline{(\frac{6}{2^3}, \frac{6}{2^3})}\}$
$\frac{7}{2^3}$	$\{\underline{(\frac{6}{2^3}, \frac{8}{2^3})}, (\frac{7}{2^3}, \frac{7}{2^3})\}$

Needless to say one can directly generate any *CF* $c \in T_n$ starting from sample (100%) and buffer (0%) droplets, i.e., using only two boundary *CF*-values. However, for many target-*CFs*, this process may require at least n mix-split steps [163]. On the other hand, our objective is to produce a *CF* from C_{min} using only one (1:1)-mixing step. In order to determine C_{min} for a given n, MTD includes two schemes: an ILP formulation and a $\sqrt{2}$-factor approximation algorithm.

7.5.1 INTEGER LINEAR PROGRAMMING (ILP) FORMULATION

It can be observed from Example 12 that for each *CF* $c_t \in T_n$ there exist several *CF*-pairs for generating c_t. Our objective is to choose a minimum-cardinality subset $C_{min} \subset T_n$, so that each $c_t \in T_n \setminus C_{min}$ can be generated from at least one pair of *CF*s (c_i, c_j), where $c_i, c_j \in C_{min}$. Note that a pair $(\frac{x}{2^n}, \frac{y}{2^n}) \in T_n$, can generate an integer *CF* if both x and y are either even or odd. In ILP formulation, we have considered, for simplicity, only those pairs that have even numbers in both of their numerators. One may also consider the *CF*s having odd numerators, but it may increase the size of C_{min}.

Let us define a Boolean variable $x_i \in \{0,1\}$ for each input concentration in $\{\frac{i}{2^n} : (0 \leq i \leq 2^n) \wedge (i \mod 2 = 0)\}$, whose *true* (1) value indicates the presence of $\frac{i}{2^n}$ in the optimal solution. Hence, the objective function is

$$Minimize : x_0 + x_2 + \ldots + x_{2^n-2} + x_{2^n} \tag{7.1}$$

Note that for each $CF = \frac{l}{2^n} \in T_n$ where $l = 1, 2, 3, \ldots, 2^n - 1$, there exist several *CF*-pairs that can be used for generating $\frac{l}{2^n}$. Let us assume that there exist k different pairs of *CF*s that can generate $\frac{l}{2^n}$. We define k Boolean variables $y_1(l), y_2(l), \ldots, y_k(l)$ for denoting the pairs that can be used for generating $\frac{l}{2^n}$ where, $y_i(l) = (\frac{i_i}{2^n}, \frac{j_i}{2^n})$. The constraints are defined as follows depending on whether l is even or odd. If l is odd then

$$\begin{aligned} x_{i_1} + x_{j_1} - 2y_1(l) &\geq 0 \\ x_{i_2} + x_{j_2} - 2y_2(l) &\geq 0 \\ &\vdots \\ x_{i_k} + x_{j_k} - 2y_k(l) &\geq 0 \end{aligned} \tag{7.2}$$

$$y_1(l) + y_2(l) + \ldots + y_k(l) \geq 1 \tag{7.3}$$

If l is even, then Expression (7.2) remains the same but the Expression (7.3) needs to be modified because $\frac{l}{2^n}$ may be present in the optimal solution. Hence, the latter expression becomes

$$y_1(l) + y_2(l) + \ldots + y_k(l) + x_l \geq 1 \tag{7.4}$$

As an illustrative example, let us consider the ILP formulation for the problem considered in Example 12. In this case, the objective function is $F = x_0 + x_2 + x_4 + x_6 + x_8$. Table 7.3 shows all source-combinations for each desired target concentration. We are considering only the pairs having an even number in their numerators. For $CF = \frac{3}{2^3}$ these pairs are: $\{(\frac{0}{2^3}, \frac{6}{2^3}), (\frac{2}{2^3}, \frac{4}{2^3})\}$. Hence we need two indicator variables $y_1(3), y_2(3)$. Here, the constraints are given by:

$$\begin{aligned} x_0 + x_6 - 2y_1(3) &\geq 0 \\ x_2 + x_4 - 2y_2(3) &\geq 0 \\ y_1(3) + y_2(3) &\geq 1 \end{aligned}$$

Similarly, for $CF = \frac{4}{2^3}$, the valid combinations are $\{(\frac{0}{2^3}, \frac{8}{2^3}), (\frac{2}{2^3}, \frac{6}{2^3}), (\frac{4}{2^3}, \frac{4}{2^3})\}$, where $(\frac{4}{2^3}, \frac{4}{2^3})$ denotes the case when $\frac{4}{2^3}$ is already selected in the optimal solution. Hence we need two indicator variables $y_1(4), y_2(4)$. The respective constraints are given by:

$$x_0 + x_8 - 2y_1(4) \geq 0$$
$$x_2 + x_6 - 2y_2(4) \geq 0$$
$$y_1(4) + y_2(4) + x_4 \geq 1$$

Analogously, we can generate constraints from other valid combinations for the remaining CFs $\frac{5}{2^3}, \frac{6}{2^3}$, and $\frac{7}{2^3}$. We then invoke an ILP-solver [64], which provides the minimum cardinality subset $C_{min} = \{\frac{0}{2^3}, \frac{2}{2^3}, \frac{6}{2^3}, \frac{8}{2^3}\}$. All other CFs in $T_3 \setminus C_{min}$ can be generated using at most one (1:1) mixing step.

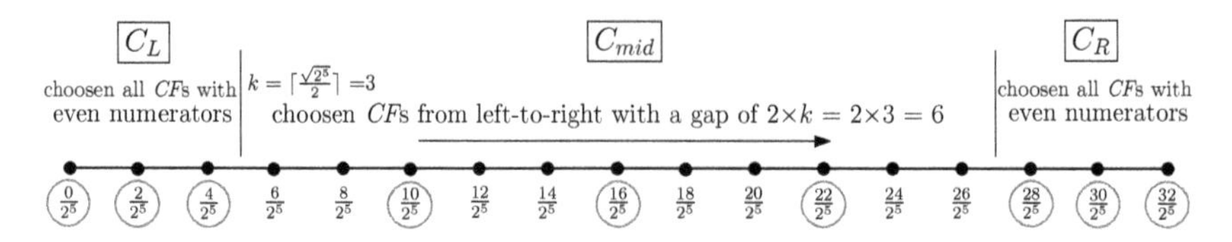

Figure 7.3: C_{approx} for $n = 5$ (with permission from IEEE [121])

7.5.2 APPROXIMATION SCHEME

The computational overhead of ILP formulation may increase rapidly with the increase of n. In order to cope with the inherent computational complexity, a constant $(\sqrt{2})$ factor approximation algorithm [168] is described below. The pseudo-code of the approximation algorithm is shown in Algorithm 6.

Algorithm 6: Approximate Source-*CF*s

Input: Depth: n
Output: $C_{approx} \subset T_n$ that generates $T_n \setminus C_{approx}$

1. $C_{approx} = \phi, N = 2^n$;
2. $k = \lceil \frac{\sqrt{N}}{2} \rceil$;
3. **for** $(i = 0; i \leq k-1; i = i+1)$ **do**
4. $C_{approx} = C_{approx} \cup \{\frac{2i}{2^n}\} \cup \{\frac{2^n - 2i}{2^n}\}$;
5. **for** $(i = 4k-2; i < N - 2k + 2; i = i + 2k)$ **do**
6. $C_{approx} = C_{approx} \cup \{\frac{i}{2^n}\}$;
7. **return** C_{approx};

Before establishing the correctness of the proposed algorithm, we illustrate it with an example. Let us assume $n = 5$ i.e., the possible CF-values are $T_5 = \{\frac{1}{2^5}, \frac{2}{2^5}, \frac{3}{2^5}, \cdots, \frac{32}{2^5}\}$. Algorithm 6 selects a set of CFs (shown as encircled in Fig. 7.3) that need to be stored in on-chip reservoirs. Assume we have arranged all CFs in

$T_5 \cup \{\frac{0}{2^5}, \frac{32}{2^5}\}$ in the increasing order of their numerators, as shown in Fig. 7.3. Algorithm 6 initially selects 3 $(= k = \lceil \frac{\sqrt{32}}{2} \rceil)$ consecutive *CF*s, that have even number in their numerators, from the left $(\{\frac{0}{2^5}, \frac{2}{2^5}, \frac{4}{2^5}\})$, and also among the right $(\{\frac{28}{2^5}, \frac{30}{2^5}, \frac{32}{2^5}\})$ from the linear arrangement of *CF*s. Then it chooses every sixth $(= 2k = 2 \times 3)$ *CF* from the set $\{\frac{10}{2^5}, \frac{11}{2^5}, \cdots, \frac{28}{2^5}\}$ and constructs $C_{approx} = \{\frac{0}{2^5}, \frac{2}{2^5}, \frac{4}{2^5}, \frac{10}{2^5}, \frac{16}{2^5}, \frac{22}{2^5}, \frac{28}{2^5}, \frac{30}{2^5}, \frac{32}{2^5}\}$. Note that for this particular example, ILP also returns the same solution, i.e., $|C_{approx}|$ is minimum. In general, Algorithm 6 may not return an optimal solution. The proof of correctness for Algorithm 6 and its approximation ratio are presented below.

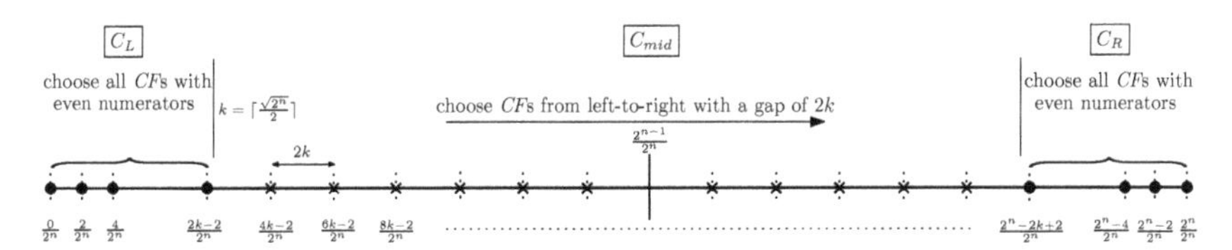

Figure 7.4: *CF*s selected by Algorithm 6 (with permission from IEEE [121])

CORRECTNESS PROOF:

We need to define some notation for proving the correctness of Algorithm 6. Let for any $n \in \mathbb{N}$, we denote $C_L = \{\frac{0}{2^n}, \frac{2}{2^n}, \ldots, \frac{2k-2}{2^n}\}$, $C_R = \{\frac{2^n-2k+2}{2^n}, \frac{2^n-2k+4}{2^n}, \ldots, \frac{2^n}{2^n}\}$ and $C_{mid} = \{\frac{2k(i+1)-2}{2^n} : i = 1, 2, \ldots, \lfloor \frac{2^n-4k+3}{2k} \rfloor\}$, where $k = \lceil \frac{\sqrt{2^n}}{2} \rceil$. Fig. 7.4 shows the selection of C_L, C_R, and C_{mid} from the entire range of *CF*s. Note that $C_{approx} = C_L \cup C_R \cup C_{mid}$. Let $C^L_{mid}(\frac{c}{2^n})$ denote the set of *CF*s produced when a $CF = \frac{c}{2^n} \in C_{mid}$ is mixed with each *CF* in C_L. Thus, $|C^L_{mid}(\frac{c}{2^n})| = k$. Similarly, $C^R_{mid}(\frac{c}{2^n})$ can be defined. Note that we can also define $C^L_R(\frac{c}{2^n})$ as the set of *CF*s generated by mixing a *CF* $\frac{c}{2^n} \in C_R$ with each *CF* in C_L. The following lemma proves an important attribute concerning $C^L_{mid}(\cdot)$.

Lemma 2 Let $\frac{x}{2^n}, \frac{y}{2^n} \in C_{mid}$, where $x \neq y$. Then $C^L_{mid}(\frac{x}{2^n}) \cap C^L_{mid}(\frac{y}{2^n}) = \phi$.

Proof It is sufficient to show that $C^L_{mid}(\frac{4k-2}{2^n}) \cap C^L_{mid}(\frac{6k-2}{2^n}) = \phi$. It can be easily verified that $C^L_{mid}(\frac{4k-2}{2^n}) = \{\frac{2k-1}{2^n}, \frac{2k}{2^n}, \ldots, \frac{3k-2}{2^n}\}$ and $C^L_{mid}(\frac{6k-2}{2^n}) = \{\frac{3k-1}{2^n}, \frac{3k}{2^n}, \ldots, \frac{4k-2}{2^n}\}$. Hence, $C^L_{mid}(\frac{4k-2}{2^n}) \cap C^L_{mid}(\frac{6k-2}{2^n}) = \phi$.

One can also prove a similar result when *CF*-values belonging to C_{mid} are combined with those in C_R. For brevity, we are only providing the statement, without giving the proof.

Lemma 3 Let $\frac{x}{2^n}, \frac{y}{2^n} \in C_{mid}$, where $x \neq y$. Then $C^R_{mid}(\frac{x}{2^n}) \cap C^R_{mid}(\frac{y}{2^n}) = \phi$.

Lemma 4 $C^L_R(\frac{2^n-2k+2}{2^n}) \cup C^L_R(\frac{2^n}{2^n})$ generates all the *CF*s within the range $[\frac{2^{n-1}-k+1}{2^n}, \frac{2^{n-1}+k-1}{2^n}]$.

Proof It can be easily verified that the set of *CF*s generated by $C_R^L(\frac{2^n-2k+2}{2^n})$ will include $\{\frac{2^{n-1}-k+1}{2^n}, \frac{2^{n-1}-k+2}{2^n}, \ldots, \frac{2^{n-1}}{2^n}\}$ and similarly $C_R^L(\frac{2^n}{2^n})$ will include $\{\frac{2^{n-1}}{2^n}, \frac{2^{n-1}+1}{2^n}, \ldots, \frac{2^{n-1}+k-1}{2^n}\}$. Hence, $C_R^L(\frac{2^n-2k+2}{2^n}) \cup C_R^L(\frac{2^n}{2^n}) = \{\frac{2^{n-1}-k+1}{2^n}, \frac{2^{n-1}-k+2}{2^n}, \ldots, \frac{2^{n-1}}{2^n}, \frac{2^{n-1}+1}{2^n}, \ldots, \frac{2^{n-1}+k-1}{2^n}\}$

The following theorem now proves the correctness of the algorithm.

Theorem 3 The set of *CF*s returned by Algorithm 6 (C_{approx}) generates all *CF*s in $T_n \setminus C_{approx}$.

Proof Let $k = \lceil \frac{\sqrt{2^n}}{2} \rceil$. Algorithm 6 initially chooses k even numbers into C_L starting from $\frac{0}{2^n}$. Note that, C_L generates $(k-1)$ odd concentrations $\{\frac{1}{2^n}, \frac{3}{2^n}, \ldots, \frac{2k-3}{2^n}\}$. Thus, C_L alone generates the set of *CF*s within the range $[\frac{1}{2^n}, \frac{2k-2}{2^n}]$. Similarly, C_R generates all *CF*s within the range $[\frac{2^n-2k+2}{2^n}, \frac{2^n}{2^n}]$. Moreover, $|C_{mid}| = \lfloor \frac{2^n+1-2(2k-1)}{2k} \rfloor = \lfloor \frac{2^n-4k+3}{2k} \rfloor$. Lemma 2 states that the set of concentrations generated by combining each concentration in C_{mid} with C_L are mutually disjoint and contiguous. So, all the *CFs* belonging to the range $[\frac{2k-1}{2^n}, \frac{2^{n-1}-k}{2^n}]$ can be generated by C_L and C_{mid}. The remaining *CF*s within the range $[\frac{2^{n-1}-k+1}{2^n}, \frac{2^{n-1}}{2^n}]$ can be generated by $C_R^L(\frac{2^n-2k+2}{2^n})$ (Lemma 4). Thus, all the *CF*s belonging to the range $[\frac{1}{2^n}, \frac{2^{n-1}-1}{2^n}]$ can be generated. Similarly, one can argue with the *CF*s belonging to $[\frac{2^{n-1}+1}{2^n}, \frac{2^n-1}{2^n}]$. Therefore, Algorithm 6 generates all *CFs* $\in T_n \setminus C_{approx}$ using only one (1:1) mixing step. Fig. 7.5 shows an example for generating the *CF*s in $T_5 \setminus C_{approx}$ from $C_L = \{\frac{0}{2^5}, \frac{2}{2^5}, \frac{4}{2^5}\}$, $C_{mid} = \{\frac{10}{2^5}, \frac{16}{2^5}, \frac{22}{2^5}\}$ and $C_R = \{\frac{28}{2^5}, \frac{30}{2^5}, \frac{32}{2^5}\}$ where $C_{approx} = C_L \cup Cmid \cup C_R$. Only the numerator ($i$) of each *CF* (= $\frac{i}{2^5}$) is shown in the Fig. 7.5. The two incoming edges emanating from node i and j to a node k represent the fact that *CF* = $\frac{k}{2^5}$ can be obtained by an (1:1) mix operation between $\frac{i}{2^5}$ and $\frac{j}{2^5}$, where $k = \frac{(i+j)}{2}$.

PERFORMANCE BOUND:

Algorithm 6 determines the output solution C_{approx} without reading the entire range of *CF*-values. In fact, it deterministically selects only a subset of *CF*-values, which is sub-linear on the size of T_n. In other words, it runs in $O(\sqrt{2^n})$ time. The following lemma proves a lower bound on the size of the optimal solution.

Lemma 5 *The minimum number of CFs required to generate a target set of size* $N = 2^n$ *is* $\lfloor \sqrt{2N} \rfloor$.

Proof Let k' be the minimum number of *CF*s required to generate any target set of size $N - k'$. In order to prove the lower bound, let us assume that (1:1) mixing of any

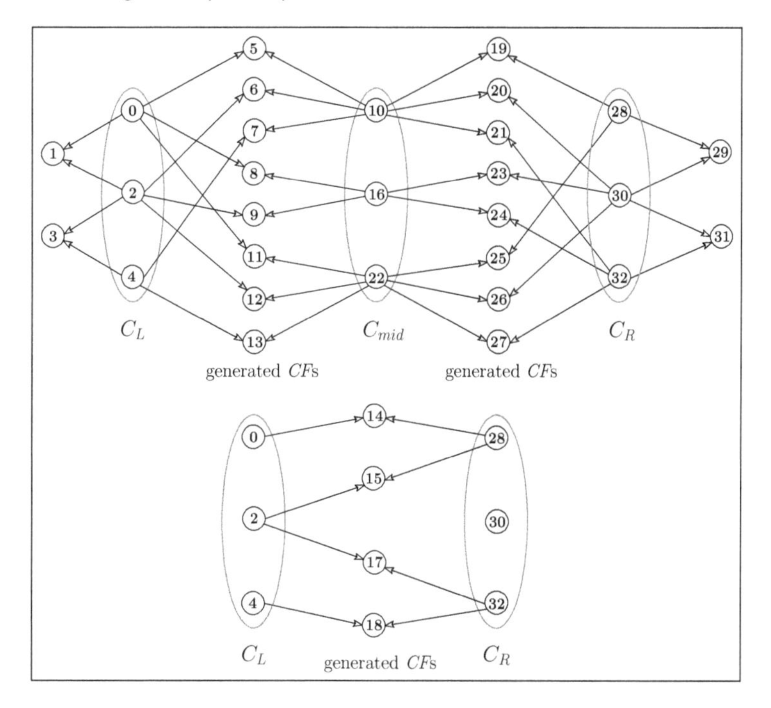

Figure 7.5: Generation of CFs in $T_5 \setminus C_{approx}$ from $C_{approx} = C_L \cup Cmid \cup C_R$ (with permission from IEEE [121])

two CFs generates a new CF. Hence,

$$\binom{k'}{2} + k' = N \Rightarrow (k')^2 + k' - 2N = 0,$$

$$\Rightarrow k' = \frac{-1 \pm \sqrt{1+8N}}{2}, N \geq 1$$

$$\Rightarrow k' = \frac{-1 + \sqrt{1+8N}}{2}, \qquad \text{(ignoring -ve solution)}$$

$$\Rightarrow k' \leq \lfloor \frac{\sqrt{8N}}{2} \rfloor = \lfloor \sqrt{2N} \rfloor$$

Theorem 4 *The approximation ratio of Algorithm 6 is $\sqrt{2}$.*

Proof Let OPT be the size of optimal solution. From Lemma 5, $OPT \geq \sqrt{2N}$. The number of CFs returned by Algorithm 6, i.e., C_{approx}, is $\leq \frac{N-2(\sqrt{N}-1)}{\sqrt{N}} + \sqrt{N} =$

$\frac{2N-2\sqrt{N}+2}{\sqrt{N}}$. Hence,

$$\begin{aligned}\text{Approximation ratio} &= \frac{|C_{approx}|}{OPT} \\ &\leq \frac{2N-2\sqrt{N}+2}{\sqrt{N}.\sqrt{2N}} \\ &= \sqrt{2} - \frac{2}{\sqrt{2N}} + \frac{\sqrt{2}}{N} \\ &= \sqrt{2} \ as\ N \to \infty\end{aligned}$$

7.6 GENERATING PARTIAL SET OF CONCENTRATION FACTORS

For certain application-specific DMFB platforms, the entire range of possible CFs (T_n) may not be necessary from the users' perspective. Instead, a subset $\mathcal{T}_n \subseteq T_n$ may be sufficient. In multiplexed bioassays, the CFs needed for different assays may be considered as the target set. Note that the ILP-based formulation can be easily adapted for determining the minimum-cardinality subset (C_{min}) of source-CFs so that any $CF \subseteq \mathcal{T}_n \setminus C_{min}$ can be produced in one (1:1) mixing step. However, it suffers from high computational overhead when n becomes large. The approximation algorithm, on the other hand, works satisfactory for the complete set of CFs (T_n) but often fares badly on a subset of T_n. In order to overcome this problem, a greedy heuristic method called Best Candidate Selection (BCS) can be deployed. Let us first describe the working principle of BCS with an illustrative example.

Example 13 Suppose we need to serve on-demand request from $\mathcal{T}_4 = \{\frac{1}{16}, \frac{3}{16}, \frac{4}{16}, \frac{7}{16}, \frac{8}{16}, \frac{9}{16}, \frac{10}{16}, \frac{11}{16}, \frac{15}{16}\}$, where $\mathcal{T}_4 \subset T_4$. The goal is to find a minimum-cardinality subset C_{bcs} consisting of source-CFs ($C_{bcs} \subset T_4$) so that any $CF \in \mathcal{T}_4 \setminus C_{bcs}$ can be produced by mixing two CFs taken from C_{bcs} in one (1:1) mixing step.

As before, C_{bcs} is initialized with the boundary CFs $\{\frac{0}{2^4}, \frac{16}{2^4}\}$. Initially, we list all CF-pairs that can be used to produce each target-CF of $\mathcal{T}_n$ in one (1:1) mixing step. Note that each target-CF of $\mathcal{T}_4$ can be generated from one or more CF-pairs. However, for few target-CFs, there may exist only a single pair of CFs. We have to include both CFs for all such target-CFs into the solution set. Since target-CF $\frac{1}{16}$ and $\frac{15}{16}$ can be generated only with the CF-pair $(\frac{0}{16}, \frac{2}{16})$ and $(\frac{14}{16}, \frac{16}{16})$, respectively, we update C_{bcs} as $\{\frac{0}{2^4}, \frac{2}{2^4}, \frac{14}{2^4}, \frac{16}{2^4}\}$. Thus each CF belonging to the subset $\{\frac{1}{16}, \frac{7}{16}, \frac{8}{16}, \frac{9}{16}, \frac{15}{16}\}$ can be produced from the updated C_{bcs} ($\{\frac{0}{2^4}, \frac{2}{2^4}, \frac{14}{2^4}, \frac{16}{2^4}\}$). Let S denote the remaining target-CFs $\{\frac{3}{2^4}, \frac{4}{2^4}, \frac{10}{2^4}, \frac{11}{2^4}\}$ as shown in Table 7.4, and $\widehat{S}$ be the union of all CFs from which S can be produced in one (1:1) mixing step. We then iteratively select CFs from $\widehat{S}$ as follows. In each iteration, a CF is selected from $\widehat{S}$ and inserted into C_{bcs}, if it maximizes the number of newly generated $CFs \in S$. Note that S and $\widehat{S}$ are updated in each iteration, since a number of newly target-CFs are produced in each iteration. It has been observed from Table 7.4 (underscored entries) that the insertion of $\frac{6}{2^4}$ into C_{bcs}

Table 7.4
Possible pairs of *CF*-values generating *CFs* $\{\frac{3}{2^4}, \frac{4}{2^4}, \frac{10}{2^4}, \frac{11}{2^4}\}$**.**

Target	Possible pairs of CF-values
$\frac{3}{2^4}$	$\{\underline{(\frac{0}{2^4}, \frac{6}{2^4})}, (\frac{1}{2^4}, \frac{5}{2^4}), (\frac{2}{2^4}, \frac{4}{2^4})\}$
$\frac{4}{2^4}$	$\{(\frac{0}{2^4}, \frac{8}{2^4}), (\frac{1}{2^4}, \frac{7}{2^4}), \underline{(\frac{2}{2^4}, \frac{6}{2^4})}, (\frac{3}{2^4}, \frac{5}{2^4})\}$
$\frac{10}{2^4}$	$\{(\frac{4}{2^4}, \frac{16}{2^4}), (\frac{5}{2^4}, \frac{15}{2^4}), \underline{(\frac{6}{2^4}, \frac{14}{2^4})}, (\frac{7}{2^4}, \frac{13}{2^4}), (\frac{8}{2^4}, \frac{12}{2^4}), (\frac{9}{2^4}, \frac{11}{2^4})\}$
$\frac{11}{2^4}$	$\{\underline{(\frac{6}{2^4}, \frac{16}{2^4})}, (\frac{7}{2^4}, \frac{15}{2^4}), (\frac{8}{2^4}, \frac{14}{2^4}), (\frac{9}{2^4}, \frac{13}{2^4}), (\frac{10}{2^4}, \frac{12}{2^4})\}$

maximizes the number of newly generated target CFs $(\frac{3}{2^4}, \frac{4}{2^4}, \frac{10}{2^4}, \frac{11}{2^4})$ of $\mathscr{T}_4$. Since $C_{bcs} = \{\frac{0}{2^4}, \frac{2}{2^4}, \frac{6}{2^4}, \frac{14}{2^4}, \frac{16}{2^4}\}$ generates all target-$CFs \in \mathscr{T}_4$ in one (1:1) mixing step, it is sufficient to store only these source-CFs into on-chip reservoirs for serving the given on-demand request profile. The outline of BCS algorithm is shown in Algorithm 7. ■

Algorithm 7: Best Candidate Selection (*BCS*)

Input: Demand target set: $\mathscr{T}_n \subseteq T_n$, depth: n
Output: $C_{bcs} \subseteq T_n$ that generates $\mathscr{T}_n$

1. $C_{bcs} = \{\frac{0}{2^n}, \frac{2^n}{2^n}\}$
2. **for** *each* $\frac{c}{2^n} \in \mathscr{T}_n$ **do**
3.   **if** $\frac{c}{2^n}$ *is generated using a single pair only* $(\frac{c_i}{2^n}, \frac{c_j}{2^n}) \in T_n \cup \{\frac{0}{2^n}, \frac{2^n}{2^n}\}$ **then**
4. $C_{bcs} = C_{bcs} \cup \{\frac{c_i}{2^n}, \frac{c_j}{2^n}\}$
5. Let $gen(C)$ be the set of all CFs that can be generated by mixing two CFs taken from C
6. $\mathscr{T}_n = \mathscr{T}_n \setminus gen(C_{bcs})$
7. **while** $\mathscr{T}_n \neq \emptyset$ **do**
8.   Let $\widehat{S} \subset T_n \cup \{\frac{0}{2^n}, \frac{2^n}{2^n}\}$ that can generate each CF of $\mathscr{T}_n$ by mixing two CFs of $\widehat{S}$ in one mixing step;
9.   Let $\frac{c}{2^n} \in \widehat{S} \setminus C_{bcs}$ be a CF, whose inclusion in C_{bcs} maximizes the size of $gen(C_{bcs} \cup \{\frac{c}{2^n}\}) \cap \mathscr{T}_n$;
10. $C_{bcs} = C_{bcs} \cup \{\frac{c}{2^n}\}$;
11. $\mathscr{T}_n = \mathscr{T}_n \setminus gen(C_{bcs})$
12. **return** C_{bcs};

Note that $\mathscr{T}_n$ is chosen as source-CFs when $|C_{bcs}| \geq |\mathscr{T}_n|$ and $|\mathscr{T}_n| \leq |C_{approx}|$. However, for $|C_{bcs}| \leq |\mathscr{T}_n|$ and $|C_{bcs}| \geq |C_{approx}|$, C_{approx} is selected as source-CFs.

7.7 REDUCTION OF ON-CHIP RESERVOIRS

In Sections 7.5 and 7.6, we have discussed how to determine the minimum-cardinality "source-*CFs*" (= # of on-chip reservoirs), which are to be stored on-chip in order to meet on-demand request of a target-*CF*, using at most one (1:1) mix-step. In this section, we show that the number of required source-*CF*s (or # of reservoirs) can be reduced by allowing increased number of mixing steps. Needless to say, in the extreme case, at least two reservoirs are needed (to store only the two boundary *CF*s = 0, 1), for which the number of mix-split operations may become n when twoWayMix algorithm [163] is used. In order to utilize such trade-off, the following strategy for successive minimization of the number of on-chip reservoirs can be adopted.

Given a set C^i_{min} of source-*CF*s, we say that a target set $T_n \setminus C^i_{min}$ is at *dilution distance* i if a target-$CF \in T_n \setminus C^i_{min}$ can be produced using at most i (1:1) mix-split steps. Thus, the target set discussed in Section 7.5 is at 1-*dilution-distance* from C_{min}. Henceforth, for convenience of notation, we will denote C_{min} as C^1_{min}.

Similarly, we may obtain 2-distance source-*CF*s (denoted as C^2_{min}) starting from the set C^1_{min} as input and applying Algorithm 7. The resulting set of source-*CF*s (C^2_{min}) can be stored on-chip so that any $CF \in T_n \setminus C^2_{min}$ can be produced in at most two mixing steps. Note that $|C^2_{min}| \leq |C^1_{min}|$.

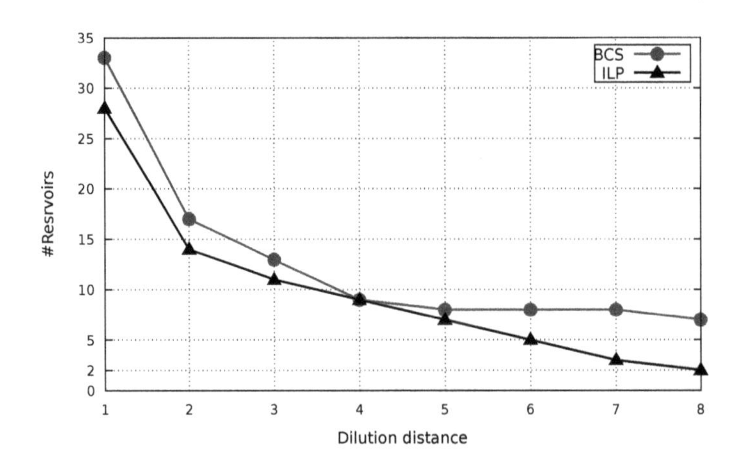

Figure 7.6: Variation of the number of on-chip reservoirs with dilution-distance for on-demand target set T_8 (with permission from IEEE [121])

Example 14 We observe that for $n = 8$, the 1-distance diluter requires 28 on-chip reservoirs for on-demand generation of any requested target-*CF* in T_8 using at most one (1:1) mixing step (Table 7.5). Fig. 7.6 shows the variation in the number of on-chip reservoirs with respect to increasing *dilution-distance*. Note that for 1 (2)-*dilution-distance*, the number of on-chip reservoirs is observed to be 28 (14) and 35 (15) when we run ILP and *BCS* algorithm, respectively. The number decreases monotonically when higher *dilution-distance* is allowed. In the extreme case when the *twoWayMix* algorithm is used for generating a target-*CF* in T_8, only two reservoirs (loaded with sample and buffer) are needed. ■

It is evident from Example 14 that reduction in the number of on-chip reservoirs has a direct consequence on sample-preparation time on-the-fly. Low-distance diluters are much more efficient since one waste droplet is generated in each mixing step. The current microfluidic technology allows to fabricate chips with a large number of built-in reservoirs. For instance, the use of 96-well chips for storing samples along with automated fluid handling systems has been reported in the literature [19, 114].

7.8 STREAMING OF DIFFERENT SOURCE CONCENTRATIONS

In previous sections, we have discussed how to choose the minimum number of "source-*CF*s" for serving on-demand request rapidly. However, one has to produce the "source-*CF*s" in advance using sample and buffer droplets. Note that existing techniques [31, 133] are not well suited for low-cost streaming of multiple *CF*s. In this section, we describe a customized Multiple Target Streaming Engine (*MTSE*) that can be used to fill the on-chip reservoirs with different "source-*CFs*" in advance. *MTSE* can be deployed to produce multiple droplets of each "source-*CF*s" using sample and buffer droplets with reduced time and cost (reactant and waste) compared to other methods [31, 133]. We present below an example for illustrating the working principle of *MTSE*.

Example 15 Assume that we need to fill two source reservoirs in advance with *CF*-values = $\frac{6}{32}$ and $\frac{10}{32}$ with 6 and 8 droplets, respectively. *MTSE* first creates an initial-forest[2] that consists of a number of singleton trees ($\tau_1, \tau_2, \tau_3 \ldots$) with "source-*CFs*". Fig. 7.7 shows the initial-forest for this example. A dilution-forest[3]

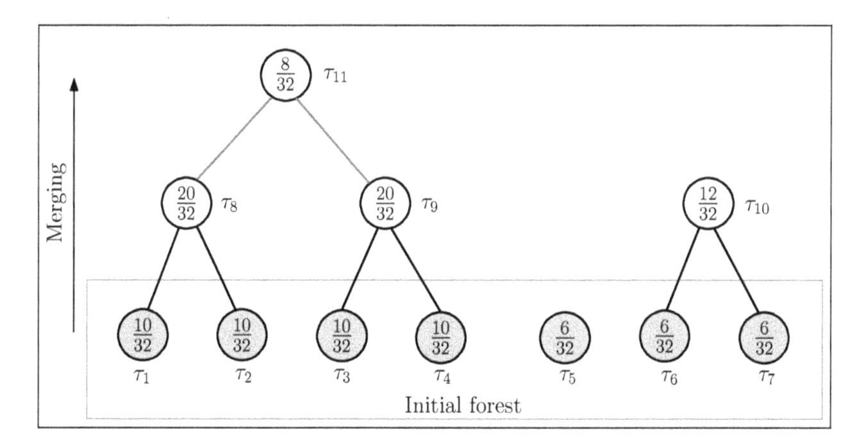

Figure 7.7: Dilution forest after initial merging (with permission from IEEE [121])

[2]collection of singleton trees in which a pink colored node represents a "source-*CF*" value.

[3]collection of trees in which each leaf node represents a source-*CF* and the root node denotes an intermediate *CF*.

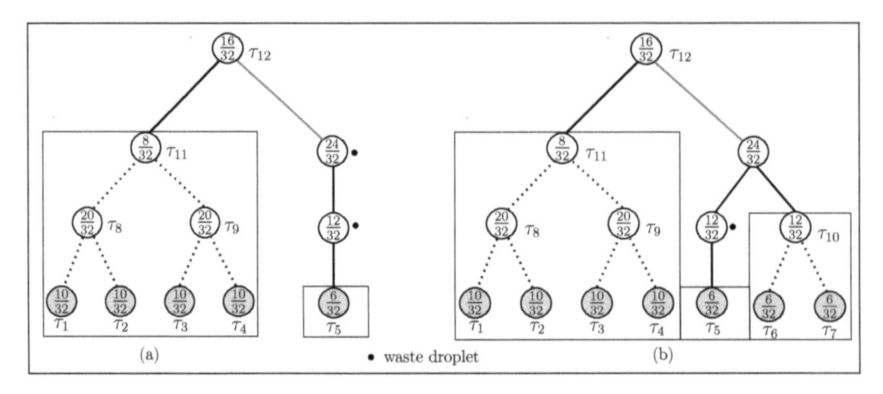

Figure 7.8: Dilution tree (a) generated by Bhattacharjee *et al.* [12] on target set $\{\frac{6}{32}, \frac{8}{32}, \frac{12}{32}\}$ (b) after waste recycling (with permission from IEEE [121])

is then constructed in bottom-up fashion from the initial-forest based on the following principle. Two singleton trees having the same *CF*-value are merged into a larger tree. This merging process is continued until one finds a pair of roots having the same *CF*-value. Note that the tree with root $CF = \frac{2^{n-1}}{2^n} = \frac{1}{2}$, cannot be merged into a larger tree. Fig. 7.7 demonstrates how a dilution-forest is created by merging the individual trees iteratively. For example, two trees having $CF = \frac{10}{32}$ can be merged into a larger tree with root $CF = \frac{20}{32}$ (τ_8). Similarly, other two trees (τ_3, τ_4) having $CF = \frac{10}{32}$ can be merged as τ_9. Finally two trees (τ_8, τ_9) with root $CF = \frac{20}{32}$ are merged into a bigger tree having root $\frac{8}{32}$ (τ_{11}).

The dilution-forest obtained following the tree-merging step consists of a number of trees with different root *CF*-values. For example, we obtain three trees with root $CF = \frac{12}{32}(\tau_{10})$, $\frac{8}{32}(\tau_{11})$, and $\frac{6}{32}(\tau_5)$ corresponding to "source-*CF*" values $\frac{6}{32}$ and $\frac{10}{32}$. In order to produce the "source-*CFs*" starting from sample and buffer droplets, one can merge the roots of these trees (τ_5, τ_{10}, and τ_{11}) with the leaves of another tree (produced by any standard sample-preparation algorithm). In doing this, we have adopted Pruning-based multiple target Dilution Algorithm (*PBDA*) [12], which creates a dilution tree for generating the desired target-*CF*-values. For example, Fig. 7.8(a) shows the dilution tree for the target-$CF = \frac{12}{32}, \frac{8}{32}, \frac{6}{32}$. In order to meet the required number of droplets for each "source-*CF*", *PBDA* attempts to reuse the waste droplets thus reducing sample-preparation cost. The final tree after waste recycling is shown in Fig. 7.8(b). Note that only one waste droplet is produced in this process. Additionally, *MTSE* also attempts to reduce the number of mix-split steps along with reactant usage. ■

7.9 EXPERIMENTAL RESULTS

In this section, we report the results of simulation experiments in detail. First, we evaluate the performance of the approximation scheme with respect to the optimal ILP-based solution. Next, we compare MTD with some state-of-the-art dilution algorithms. We have implemented the simulation platform in Python and used IBM ILOG CPLEX [64] optimization tool for solving the ILP problem. All experiments were performed on a 2 GHz 64-bit Linux machine with 8 GB memory.

Table 7.5
Performance for the complete set of *CF*s in T_n.

	ILP		Approximation		*BCS*	
T_n	$\|C_{min}\|$	Time (sec.)	$\|C_{approx}\|$	Time (sec.)	$\|C_{bcs}\|$	Time (sec.)
T_3	4	0.0×10^{-2}	4	0000	4	0.002
T_4	6	1.0×10^{-2}	6	0.002	6	0.006
T_5	9	3.0×10^{-2}	9	0.002	11	0.038
T_6	13	50×10^{-2}	14	0.003	16	0.387
T_7	19	$2.2\times10^{+3}$	20	0.004	24	4.460
T_8	28	$3.5\times10^{+4}$	30	0.009	35	53.39
T_9	53[a]	$4.3\times10^{+5}$	43	0.020	50	640.1
T_{10}	73[a]	$6.0\times10^{+5}$	62	0.040	76	8125

[a] ILP did not terminate for these cases

7.9.1 PERFORMANCE EVALUATION OF ILP AND THE APPROXIMATION SCHEME

In order to analyze the performance of the ILP solution, the approximation scheme, and the heuristic method (*BCS*), we have considered the target set T_n of size $2^n - 1$ for $n = 3, 4, \ldots, 10$. It has been observed that the ILP-solver terminated successfully up to the values of n = 8 while determining the optimal number of intermediate concentrations (C_{min}). Unfortunately, for $n \geq 9$, it did not terminate in reasonable time (see Table 7.5). On the other hand, both the approximation scheme and *BCS* method produce a solution quickly for $n \leq 10$. The size of the output set ($|C_{min}|$) and the corresponding CPU-time are reported in Table 7.5 for different values of n. Note that the approximation scheme produces good solutions for all practical scenarios satisfactorily.

7.9.2 PERFORMANCE EVALUATION OF BCS SCHEME

In this subsection, we report the performance of the *BCS* algorithm considering randomly-generated target sets of different sizes. We have computed C_{bcs} (C_{min}) and CPU-time for the *BCS* (ILP) method. Comparative results are presented in Table 7.6. Note that the ILP-based method requires a large amount of time even for a small-sized target set (30); for large target sets (100), ILP does not terminate (thus may produce a non-optimal solution). *BCS* algorithm, on the other hand, efficiently generates a solution in reasonable amount of time for each case.

Table 7.6
Performance for the partial set of *CFs* in $\mathscr{T}_8$.

	ILP		*BCS*	
$\lvert\mathscr{T}_8\rvert$	$\lvert C_{min}\rvert$	Time (sec.)	$\lvert C_{bcs}\rvert$	Time (sec.)
30	15	8696.29	16	3.32
50	18	9393.75	18	4.93
100	24[a]	7790.51	25	8.31

[a] ILP did not terminate

7.9.3 PERFORMANCE EVALUATION OF MTSE

We conduct further experiments for evaluating the efficiency of the *MTSE* algorithm. In our experiments, we select the "source-*CF*s" for accuracy level (n) = 8 (as obtained by the ILP-solver). Demands for each "source-*CF*" are assumed to be 2, 4, 8, 16, and 32. We calculate the mean value of mix-split steps, sample (buffer) droplets and waste droplets required by *MTSE*. For calculating the mean value, we divide the total count of a particular parameter by the number of demand droplets. We observe that the average cost per target-droplet (i.e., the number of mix-split steps, sample- and buffer-droplets needed for producing source-*CF*s divided by demand-value) decreases noticeably until the demand-value reaches 8 and after that, it becomes almost constant. Note that *MTSE* produces negligible amount of waste in each case, and for demand-value = 16 and 32, no waste droplet is produced (see Fig. 7.9).

7.9.4 PERFORMANCE OF THE INTEGRATED DILUTION SCHEME WITH MTSE

We evaluate the overall performance of the on-demand dilution algorithm including the cost required to fill the on-chip reservoirs using *MTSE* algorithm. We compare one-distance dilutor with two standard algorithms *RSM* [57], and *WARA-sharing-only* [100]. In our experiments, we have chosen target sets of different sizes ($N =$

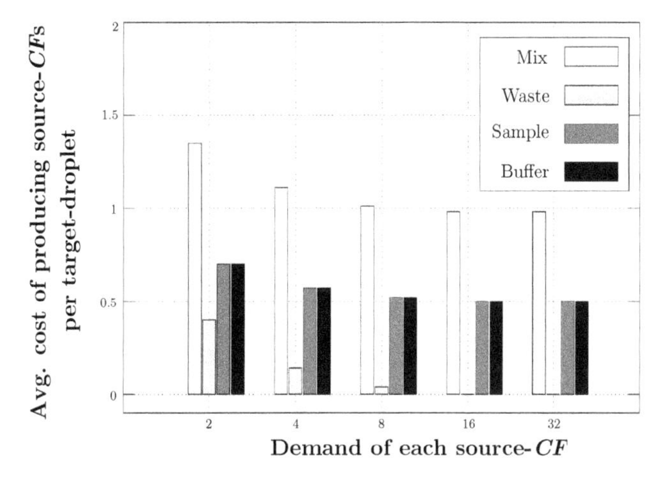

Figure 7.9: Average number of mix steps, waste, sample, and buffer consumption in *MTSE* (with permission from IEEE [121])

10, 20, 30, 40, 50) for $n = 8$. Additionally, for a fixed-size target set, we have chosen 50 random instances. We compute the mean value (μ), standard deviation (σ) of mix-split steps ($\bar{n}_m$), the number of waste droplets ($\bar{n}_w$), the number of sample ($\bar{n}_s$) and buffer ($\bar{n}_b$) droplets needed by *RSM* [57], *WARA-sharing-only*[4] [100], and by MTD. For computing the mean value μ, we divide the total count of a particular parameter by the number of different target sets. As evident from Fig. 7.11, MTD

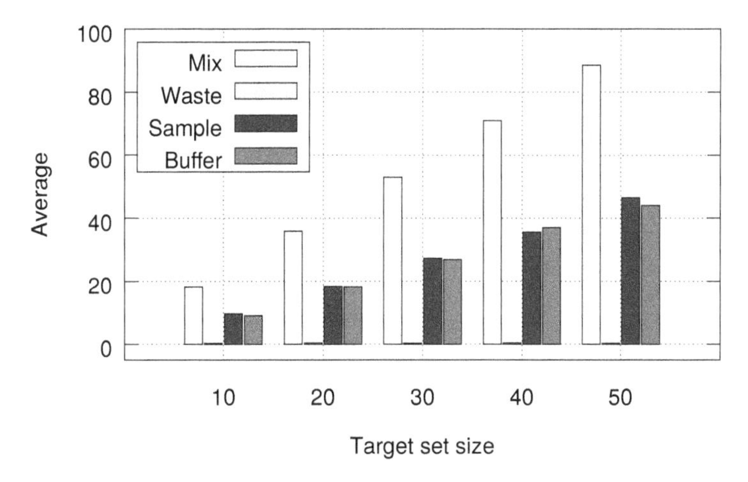

Figure 7.10: Average mix, waste, sample, and buffer consumption in *MTSE* (with permission from IEEE [121])

[4]only droplet sharing part has been implemented.

outperforms both *RSM* and *WARA* significantly in terms of reactant cost (sample and buffer), and consequently, in waste production. However, it requires more mix-split steps compared to *RSM* when the size of the target set size is 50. Note that on-demand service-time (the time required to prepare a target-*CF*) for MTDs involves only one (1:1) mix operation when asked for, as the source-reservoirs are filled ahead of time. We also evaluate the performance of *MTSE*, which is used to fill the on-chip reservoirs with source-*CFs*. Fig. 7.10 shows the average $\bar{n}_m, \bar{n}_w, \bar{n}_s$ and $\bar{n}_b$ considering the cumulative demand for each source-*CF*.

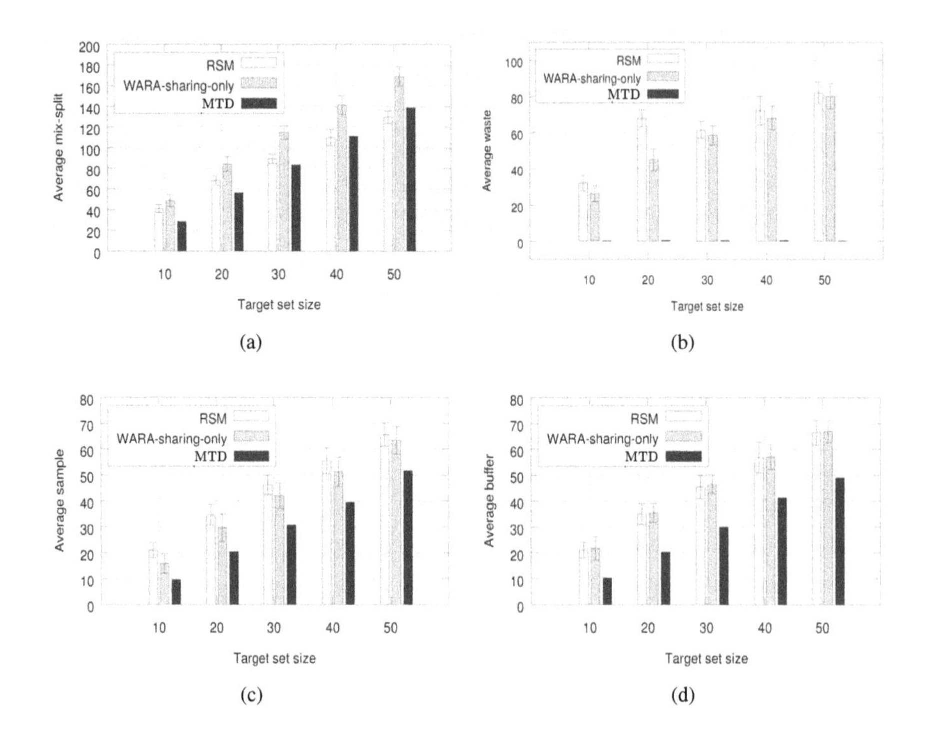

Figure 7.11: Average (a) mix-split ($\bar{n}_m$), (b) waste ($\bar{n}_w$), (c) sample consumption ($\bar{n}_s$), and (d) buffer consumption ($\bar{n}_b$) for all four methods over different target-set sizes when $n = 8$ (with permission from IEEE [121])

7.10 ERROR-OBLIVIOUSNESS

MTD is essentially a two-phase process. In the first phase, it selects the smallest set of source-*CF*s using an ILP/approximation algorithm for the entire range of *CF*s, or using the BCS method for a partial set of *CF*s. In the second phase, MTSEs creates a dilution-tree from which the desired source-*CF*s can be generated in advance. Both

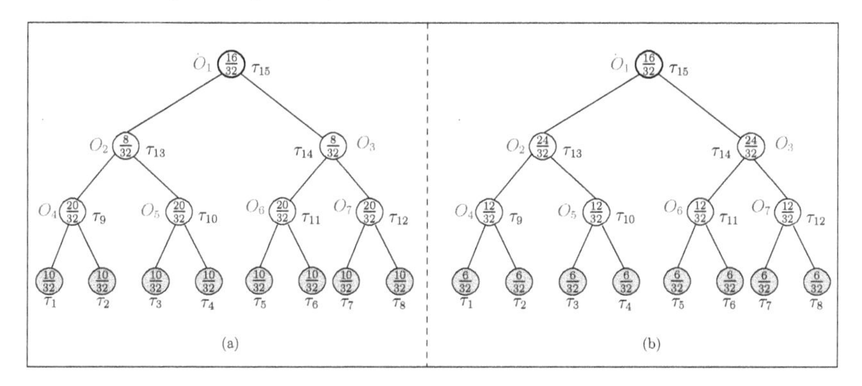

Figure 7.12: Dilution tree generated by the baseline approach (presented in Chapter 6) for source-*CF*s = $\frac{6}{32}$ and $\frac{10}{32}$ (with permission from IEEE [121])

methods work well in ideal scenario, i.e., in the absence of any volumetric split-error. However, in an error-prone environment, they require certain modification in order to serve dynamically-arriving requests (target-*CF*s), reliably. Error-tolerance can be achieved by making the mix-split steps of the dilution-tree constructed by the MTSE algorithm, insensitive to any volumetric split-errors. In order to achieve this, we adopt the baseline approach (discussed in Section 6.4) for generating the "source-*CF*" starting from sample and buffer droplets, instead of using MTSE. Note that no waste-droplets is produced when the baseline algorithm is executed. The second phase is intrinsically reliable. To produce reliable target-droplets, we use a few on-chip reservoirs (for temporarily storing intermediate droplets) [124] or a k-ary rotary-mixer[5] [131] (if we want to avoid unloading of droplets), while generating the target-*CF*s from the set of source-*CF*s. In general, a smaller value of dilution-distance is preferable since sample-preparation time increases when dilution-distance increases. An example for generating the source-*CF*s = $\frac{6}{32}$ and $\frac{10}{32}$ with the baseline approach, is shown in Fig. 7.12. It has been observed that it utilizes all intermediate droplets to generate the target-droplets without producing any waste. Since for a given target-*CF*, all daughter-droplets obtained at the leaf-nodes of the baseline tree are mixed together, volumetric split-errors, if any, are finally canceled.

7.11 CONCLUSIONS

In this chapter, we have addressed the problem associated with the generation of multiple target concentrations of a sample on a DMFB platform, where the information regarding the concentration profile is not known in advance, i.e., when the demand vector arrives on-the-fly depending on users' requirement. The optimization of sample-preparation cost and time becomes more complex in such a scenario. In MTD, this problem is solved by identifying a small subset of concentration

[5] A k-ary rotary mixer can efficiently generates $2k$X-volume target solution with just one ($k:k$) mixing operation.

factors (source-*CF*s) *a-priori*, which can be used to generate any other target-*CF* by executing only a few mix operations. In other words, MTD reduces the problem of handling a dynamic subset of *CF*s to the selection of a pre-computed static subset. Several techniques based on ILP, approximation algorithm, or heuristics can be used to reduce the size of this static subset of *CF*s. In addition, a cost-effective droplet-streaming engine (MTSE) that produces the required source-*CF*s is described. The static subset of *CF*s can be stored in on-chip reservoirs and their number (i.e., the cardinality of the subset) can be traded-off with sample-preparation time. Finally, an error-tolerant mechanism for producing target droplets with desired accuracy is discussed. As future research work, one may further study the issues in on-demand mixture preparation with multiple reagents, and the optimization of bio-protocols that are fed to reactors in real-time [68, 69].

8 Robust Multi-Target Sample Preparation with MEDA Biochips

Over the last few decades, droplet-based digital microfluidic technology has emerged as a promising platform in the healthcare industry for enabling point-of-care (PoC) clinical diagnosis and automating biochemical assays with various applications to genetic engineering, synthetic biology, DNA analysis, and drug discovery, among others [39]. Compared to the early generation of continuous-flow microfluidics (where fluids flow through micro-channels), digital microfluidic biochips (DMFBs), wherein fluids are manipulated as discrete-volume droplets on a 2D-array of electrodes, offer a more flexible architecture, better reconfigurability of resources, and ease of interfacing with other technologies. By applying a programmed sequence of control signals to the individual electrodes, nano/pico-liter sized droplets can be electrically actuated to perform basic fluidic operations such as dispensing, transporting, mixing, and splitting of droplets. These coin-sized chips offer more convenience in terms of high throughput, portability, automation, low reagent-consumption, fast reaction-time, and are likely to replace expensive biochemical bench-top procedures currently used in hospitals, research labs, or at PoC stations. Recently, an improved version of DMFBs that are implemented as micro-electrode-dot-array (MEDA) has been proposed to achieve full scalability, granularity, and reconfigurability of droplet management [177, 178]. Sample preparation, which primarily involves dilution and mixing of fluids in certain ratios, is an important preprocessing step that is needed in almost all bioprotocols [40, 43, 163]. The optimization goal of sample-preparation algorithms is to minimize one or more of the following parameters: (i) the costly reactant-usage [60], (ii) waste-droplet emission [131], (iii) sample-preparation time, i.e., the number of mix-split operations [163], and (iv) the impact of fluidic errors [122].

MEDA-based digital microfluidic biochips recently have emerged for implementing a wide range of bioprotocols including error-correction and sample preparation on handheld devices [78, 92–94, 194]. MEDA-biochips are typically equipped with thousands of microelectrodes each having in-built actuation/sensing circuit [95]. Because of the high device density, they are more prone to hardware and functional defects, which may impact the execution of bioassays [96]. For example, the volume of daughter-droplets following a split-operation may change unexpectedly due to non-uniform electrode coating, charge trapping, or electrode breakdown. As a result, during sample preparation, errors in the target-*CF* may increase noticeably. This is a severe problem, since the outcome of the assay must be dependable in

DOI: 10.1201/9781003219651-8

many biomedical applications such as in pathological diagnostics, forensics, DNA analysis, and drug design. The management of such errors thus becomes more crucial in life-critical applications where the precision of the assay-outcome cannot be compromised. Therefore, efficient error-recovery techniques are required for sample preparation from the perspective of reliability and robustness.

In the last few years, a number of error-recovery strategies have been proposed for DMFBs [3,4,35,58,96,105,106,122,140,192] to manage such undesirable errors. Most of the permanent hardware defects in DMFBs such as electrode-shorts, electrode stuck-at faults, insulation breakdown can be detected by conducting either offline structural or routing tests [111,190], or online tests [112]. However, transient and functional defects such as electrode contamination/degradation, charge-storage, or errors due to unbalanced droplet-splitting, inaccurate dispensing or delayed droplet transport require online error-detection using sensing and cyber-physical recovery mechanisms [105, 192]. The main idea behind online error-recovery is as follows: During the execution of the bioassay, intermediate droplets are sent to designated checkpoints in order to verify their correctness by an on-chip sensor. If an error is detected, the droplet is discarded as waste. After that, a portion of the assay is re-executed by rolling back to the previous checkpoint in order to produce a correct droplet.

While the above approach addresses the problem of split-errors, a rollback scheme inherits a number of drawbacks, such as (i) it is costly in space and time [35, 58, 105, 106, 192], (ii) cannot tolerable dispensing errors, (iii) it fails to provide any bound on the number of rollback iterations, thus assay-completion time remains unknown, (iv) cannot complete error-recovery operations when not enough copy-droplets are available, and (iv) it is unable to provide solution when errors are introduced by imperfect split-operations [122]. Recently, Poddar *et al.* proposed a "roll-forward" error-correcting technique [122], and later an error-oblivious method [124] for reliable sample preparation with DMFBs. However, it increases sample-preparation cost due to certain architectural constraints of these chips. Besides that, all prior sample-preparation methods for error-correction focused on producing only a single target *CF* and without considering any dispensing errors while obtaining droplets from a reservoir. Besides reliability, the generation of multi-target samples is needed in many real-life applications [72, 76, 121, 133, 149, 185]. In the titration process, a series of dilutions of a biological reagent such as a linear, harmonic, logarithmic, Gaussian or parabolic are often required [72].

In this chapter, we describe a solution methodology introduced by Poddar et al. [119] that can be used for accurate multi-*CF* sample preparation on MEDA platforms. In order to guarantee the robustness of the procedure, MEDA-specific droplet operations [176] are deployed during the execution of the dilution assay. The method is referred to as MTM (Robust Multi-Target Sample Preparation with MEDA), the characteristics and scope of which in contrast to prior approaches are summarized in Table 8.1. The highlights of MTM are listed below:

- MTM implements error-tolerant sample preparation using only "aliquoting[1]-and-mix"[2] with differential-size aliquots, obviating the need for using traditional "mix-split" scheme[3]. This way, the main source of errors (volumetric split-error due to unbalanced splitting) is completely eliminated, and no sensing is needed while producing the target-*CF*. Moreover, since splitting of a merged droplet is not needed, it neither produces any waste droplet nor allows an imperfect split to occur.
- MTM also provides a solution for handling dispensing errors during sample preparation.
- No extra time is needed for volumetric-error recovery or sensing operations.
- MTM provides an additional advantage of producing accurate multiple dilutions of a sample in a fully parallelized and cost-effective fashion without using any intermediate droplet-sharing, a technique, which was commonly used to generate multiple target-*CF*s [100, 113]. Because of precedences among shared droplets, previous multi-target methods are neither conducive to fault-tolerance nor suitable for parallelization.

The rest of the chapter is structured as follows. Preliminaries and background of MTM are described in Section 8.1. Section 8.2 sketches the underlying idea. The effect of dispensing error is shown in Section 8.3. The problem is formally described in Section 8.4. Section 8.5 discusses the resulting methodology for reliable sample preparation with MEDA-biochips. Subsequently, the overall error-tolerance mechanism is presented in Section 8.6. Reliable generation of multi-target dilutions of a sample is discussed in Section 8.7. Experimental results are summarized in Section 8.8. Conclusions and discussion on future work appear in Section 8.9.

8.1 PRELIMINARIES AND BACKGROUND

8.1.1 DIGITAL MICROFLUIDICS WITH MEDA

The architectural details of Micro-Electrode Dot Array (MEDA) chips and their use in various error-recovery methods have been discussed earlier in Chapter 1 and Chapter 3. Due to the architectural flexibility, MEDA biochips can perform some advanced microfluidic operations, e.g., channel dispensing and fluid merging operation [79, 97, 176], which are not so convenient in traditional DMFBs. MEDA can precisely dispense a certain amount of fluid via channel dispensing operation. For example, it can dispense two droplets of volumes $\frac{1}{2}$X and 2X, from a 4X-volume source droplet (see Fig. 8.1). Note that from the viewpoint of underlying principle, channel dispensing operation is different from splitting operation.

---

[1] The granularity of MEDA-chips enables processing of droplets with fractional volumes, which are called aliquots.

[2] Two or more aliqout-droplets (with same or different volumes) are mixed; the mixed droplet is used to perform the subsequent mix-operations.

[3] Two unit-volume droplets are mixed followed by balanced splitting to produce two daughter-droplets. One of them is used to perform a following mix-operation, and the other is either discarded as waste or stored for subsequent use.

Table 8.1
Comparative features of MTM [119] against prior art.

Method	Waste-free?	Underlying Architecture?	Split-less?	Error-tolerant?	Consider dispensing errors?	Reliable for multiple errors?	Verify droplet results?	System responsiveness?	Applicable for multiple targets*?	Error-recovery steps?
[3, 4, 58, 105, 106, 192]	No	DMFB	No	Yes	No	No	Yes	Low	No	Indefinite
[122]	No	DMFB	No	Yes	No	Yes	Yes	Low	No	Definite
[124]	No	DMFB	No	Yes	No	Yes	Yes	High	No	Definite
[195]	No	DMFB	No	Yes	No	No	No	High	No	Definite
[94, 193]	No	MEDA	No	Yes	No	No	Yes	Medium	No	Indefinite
[92]	No	MEDA	No	Yes	No	No	Yes	Medium	Yes	Indefinite
MTM [119]	Yes	MEDA	Yes	Yes	Yes	Yes	No	High	Yes	Definite

*Will produce all target-*CF*s at the same time in parallel

Besides that, MEDA biochips can merge various volumes of droplets into a single droplet through fluid-merging operation and mix them thoroughly using an advanced lamination mixer module [176]. For example, it can mix two droplets of volumes 2X and 4X, following a fluid-merging operation. In lamination mixing, a merged droplet is repeatedly split and recombined in opposite directions to expedite the mixing process. However, it has been experimentally shown that there is a minimum-volume constraint associated with MEDA-droplets, which is called an "aliquot" [97].

Most of the prior work on sample preparation with conventional DMFBs use the (1:1) mixing model for mix-split operations. However, the inherent architecture of MEDA-chips is capable of supporting more generalized $(k:l)$ mixing models $(k,l \in \mathbb{N})$. In the $(k:l)$ mixing model, mixing is performed between k micro-droplets of CFs = C_1 and l micro-droplets of CFs = C_2. After mixing, a balanced split operation produces two droplets of $CF = \frac{(k \times C_1 + l \times C_2)}{(k+l)}$, each having $\frac{(k+l)}{2}$ micro-droplets.

8.1.2 SPLIT-ERROR

Split errors have a significant effect on the accuracy of the sample preparation. In fact, a volumetric split-error of ε $(0 \leq \varepsilon < 1)$ produces two daughter droplets of volume $(1+\varepsilon)$ and $(1\text{-}\varepsilon)$. When such an erroneous droplet is used in a subsequent

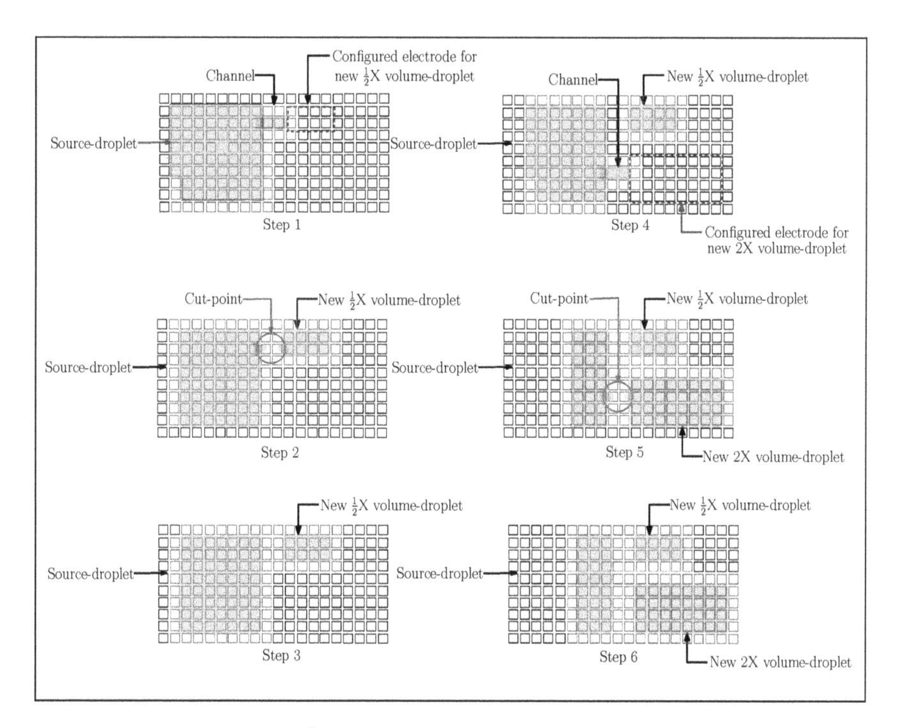

Figure 8.1: Dispensing of $\frac{1}{2}$X- and 2X-volume droplets from 4X-volume source droplet via channel dispensing operation (with permission from ACM [119])

mix-split step, the *CF* of the target-droplet may change unexpectedly [163]. The following example illustrates the problem.

Example 16 Consider the mixing tree as shown in Fig. 8.2, which ideally produces a target-$CF = \frac{71}{128}$ with accuracy-level $n = 7$. Furthermore, assume that a single volumetric split-error with $\varepsilon = 0.07$ [129] occurs at the mix-split operation O_5. This changes the expected *CF*-value from $\frac{71}{128}$ to $\frac{70.26}{128}$ and, hence, introduces a *CF*-error of $(\frac{-0.74}{128})$. This error exceeds the allowable error-limit of $(\pm\frac{0.5}{128})$ and, hence, yields, a wrong result [122, 124].

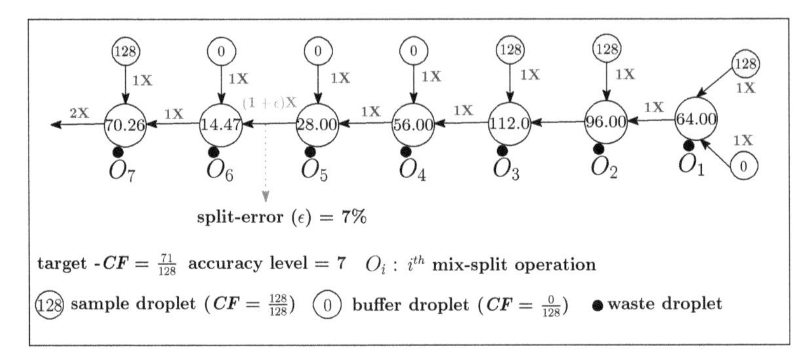

Figure 8.2: Effect of a single volumetric split-error on target-$CF = \frac{71}{128}$ for accuracy level 7 (with permission from ACM [119])

In addition, the effects of volumetric split-errors on the target-*CF*s are shown in Table 8.2 when they are prepared following classical methods such as TWM [163] and REMIA [60]. The erroneous concentration-values $(\overline{CF})$ (Column 4 of Table 8.2) obtained by both methods are shown in Table 8.2 for each target-*CF* (Column 2) when one or more volumetric split-errors (Column 3) occur in the mixing path. Note that the *CF*-error (Column 5) exceeds the allowable error-tolerance limit $(\pm\frac{0.5}{2^7})$ in every case – even for a single volumetric split-error. Furthermore, the *CF*-error in the target-droplet increases rapidly when multiple errors occur in the mixing path [124]. Hence, the reliability of the target-*CF* strongly depends on the outcome of split operations during sample preparation. Unfortunately, an ideal split operation is hard to guarantee from a real-life perspective.

8.1.3 DISPENSING-ERROR

Although MEDA biochips can generate droplets with different volumes in controlled fashion, volumetric-errors may occur during dispensing operations due to the application of unequal voltages accross the cut-point. Typically, the magnitude of the dispensing error on MEDA biochip is less than $\pm 1\%$ [181, 193], which may still affect the desired target-*CF* badly. For example, *CF*-error in the target-droplet becomes $\frac{0.58}{128}$ because of dispensing errors, as shown in Fig. 8.3, thus exceeding the

Table 8.2
Impact of volumetric split-errors to target-*CF*s for accuracy 7.

Method	*CF*	erroneous-steps	$\overline{CF}$	*CF*-error	*CF*-error< $\frac{0.5}{128}$?
BS [163]	$\frac{21}{128}$	[3,5,6]	$\frac{19.73}{128}$	$\frac{1.27}{128}$	no
	$\frac{81}{128}$	[6]	$\frac{79.49}{128}$	$\frac{1.59}{128}$	no
	$\frac{113}{128}$	[4, 7]	$\frac{111.99}{128}$	$\frac{1.01}{128}$	no
REMIA* [60]	$\frac{47}{128}$	[3]	$\frac{46.42}{128}$	$\frac{0.58}{128}$	no
	$\frac{71}{128}$	[6]	$\frac{70.25}{128}$	$\frac{0.75}{128}$	no
	$\frac{106}{128}$	[2,3]	$\frac{104.09}{128}$	$\frac{1.91}{128}$	no

$\overline{CF}$: generated *CF*; *CF*-error: $|CF - \overline{CF}|$; *considering error only at interpolated dilution phase.

error-tolerance limit $\pm\frac{0.5}{2^7}$. Therefore, a mechanism for tolerating dispensing errors during sample preparation is also essential in order to strengthen reliability.

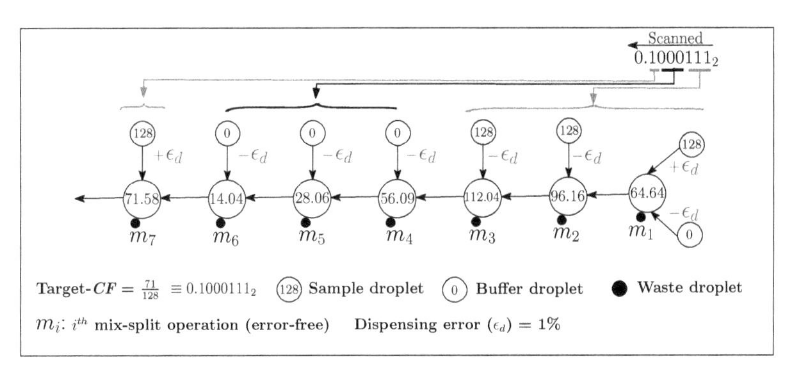

Figure 8.3: Effect of dispensing errors on target $CF = \frac{71}{128}$ using TWM [163] on DMFBs (with permission from ACM [119])

8.1.4 MINIMIZATION OF WASTE DROPLETS

Minimization of waste-droplets is one of the crucial steps in biochemical assays because management of such spare droplets is a challenging and expensive task. Waste-droplets may require separate reservoirs and extra time for routing, which in turn, may further contaminate a number of electrodes along their routing-paths. Although a large number of sample-preparation algorithms were proposed in the last few years, they mainly focused on the minimization of the number of mix-split steps and cost of reactant usage [57, 60, 131, 163]. In some other approaches [100, 135], some of the waste-droplets produced in the intermediate mix-split steps are used in the assay to reduce cost.

8.2 MTM: MAIN IDEA

MTM provides an alternative strategy for multi-target sample preparation, which is reliable (both against split-error and dispensing error) and waste-free. The main idea is to eliminate the step, which is the main source of errors, namely the splitting operation following a merge. Moreover, if no splitting occurs, volumetric errors simply do not appear anymore and sample preparation becomes significantly more reliable[4]. However, at the same time, conducting sample preparation without splitting, doubles the volume of merged droplets after each mix operation; this eventually would lead to a resulting target droplet of exponential volume 2^nX (with n being the depth of the corresponding dilution-tree). This depth should be kept as small as possible.

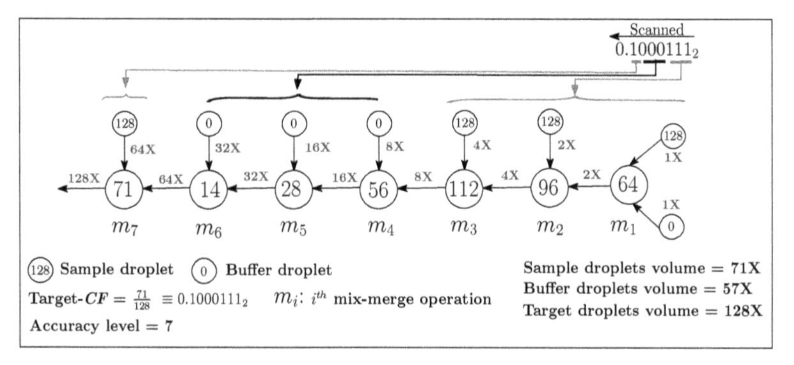

Figure 8.4: Generation of error-free target droplet with $CF = \frac{71}{128}$ using *TWM* [163] on DMFBs (with permission from ACM [119])

In twoWayMix (TWM) [163], a *skewed dilution-tree*[5] of minimum depth is constructed, under the (1:1) mixing model [100]. Here, the target-*CF* is first represented as a binary string and then the bits are scanned from right-to-left. Depending on the current bit value (1 or 0), the method then mixes a sample droplet (for 1) or buffer droplet (for 0) with the current droplet. The following example illustrates the idea.

Example 17 Consider the problem of producing target-$CF = \frac{71}{128}$ using an TWM-based approach, but without performing any split-operation. Fig. 8.4 shows the corresponding dilution tree. Initially, a sample droplet and a buffer droplet is mixed according to the least-significant 1-bit in the binary representation of the target-*CF* (1000111_2). Following the next two scanned bits from right-to-left, *twoWayMix* mixes sample droplets with the current intermediate *CF*s in next three mix-split operations. After that buffer droplets are added in subsequent three mix-split operations.

[4]by assuming that the dispensing operations are reliable. Later, in Section 8.3, we will show how dispensing errors can be tolerated.

[5]A binary tree whose depth is in the order of the number of nodes.

Finally, it generates the target-*CF* by mixing a sample-droplet with the current intermediate *CF* ($\frac{14}{128}$) in the last mix-split operation.

Needless to say, this scheme requires manipulation of droplets with substantial sizes (the volumes of the input sample/buffer droplet in each "merge-mix" step are shown in Fig. 8.4 in blue, whereas the volumes of intermediate droplets are shown in red). Overall, it can be seen that when the desired target-ratio is achieved, the volume of the final target droplet becomes 128X. Performing mixing operations with such large droplets is a complex task in traditional DMFBs and thus, dedicated mixer modules are required to enable such operations. Clearly, this limits the benefits of reconfigurability supported by DMFBs and increases the cost of the biochip further. However, this problem can be circumvented by adapting a scheme akin to TWM on MEDA-biochips without any split-operation by using the power of "droplet-aliquoting" [193] which can be implemented conveniently on a MEDA chip.

As reviewed in Chapter 1, MEDA is an advanced version of conventional DMFBs which allows accurate and flexible control of droplet volumes as well as shapes in a fine-grained manner. For example, in the MEDA-architecture shown in Fig. 1.4, we assume that a 1X-size droplet on a conventional DMFB covers the area of (4 $\times$ 4) sub-array on a MEDA-chip, i.e., it covers 16 microelectrodes. Hence, aliquots or micro-droplets of different sizes such as 1X, $\frac{15}{16}$X, $\frac{14}{16}$X, $\cdots$, $\frac{2}{16}$X, $\frac{1}{16}$X, can be formed on MEDA. Efficient manipulation of aliqouts on MEDA-architecture has been demonstrated elsewhere [95, 96, 176, 178]. Overall, this capability allows to address the above drawback of DMFBs, as illustrated in the following example:

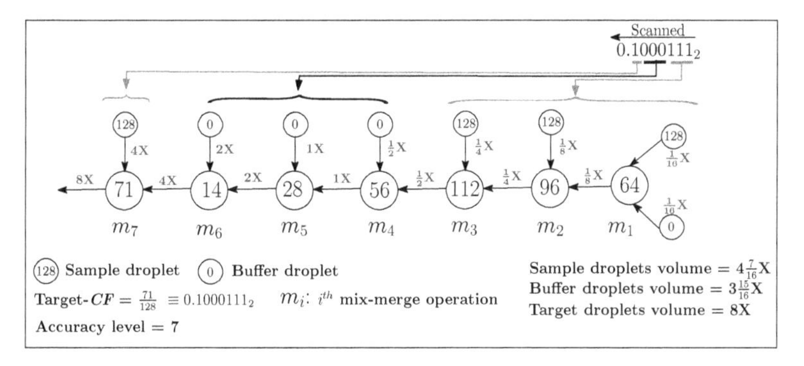

Figure 8.5: Generation of target-*CF* = $\frac{71}{128}$ using *TWM* on MEDA biochips without any split operation (with permission from ACM [119])

Example 18 An example for producing the target-*CF* = $\frac{71}{128}$ using TWM on MEDA biochips is shown in Fig. 8.5. It can be seen that the proposed "split-less" method consumes $\frac{71}{16}$X samples, $\frac{57}{16}$X buffers and requires 127-unit mixing-time[6]. Note that it does not produce any intermediate waste droplet. However, it produces 8X-volume

[6]Mixing operations are performed between micro-volume MEDA droplets. We assume that two micro-volume MEDA droplets require 1-unit of mixing time.

target-droplets and consumes a substantial amount of reactant droplets. Although the method does not need any splitting, the volume of the target droplet is increased significantly, which may not be always desirable. In this context, if the minimum allowable droplet volume on the biochip is 1X, and the required target-volume is 8X with $CF = \frac{71}{128}$, then it cannot be produced without generating any intermediate waste-droplet starting from $CF = \frac{0}{128}$ and $CF = \frac{128}{128}$. In MTM, it is implicitly assumed that target droplets that exceed user's requirement are not designated as waste-droplets; only the unused "intermediate droplets" are truly waste-droplets, as they are neither asked by the user, nor utilized elsewhere in producing other target droplets.

In order to avoid these drawbacks, an Integer Linear Programming (ILP)-based method is described next to realize a given target-*CF* of a sample using exponential *CF*-values (ECFs) [100] of the same sample (i.e., $CF = \frac{1}{2^x}$, for positive integers x), as inputs, and without performing any split operation. The ILP-method minimizes the cost of sample preparation by minimizing the volume of the target-droplet. In order to supply the required droplets with ECFs, MTM initially creates a dilution-forest, based on which they are generated and stored in the on-chip reservoirs.

Integer Linear Programming (ILP) formulation: The ILP-model uses a dilution tree whose structure is shown in Fig. 8.6. It utilizes different ECFs for producing a particular target-*CF*, as follows:

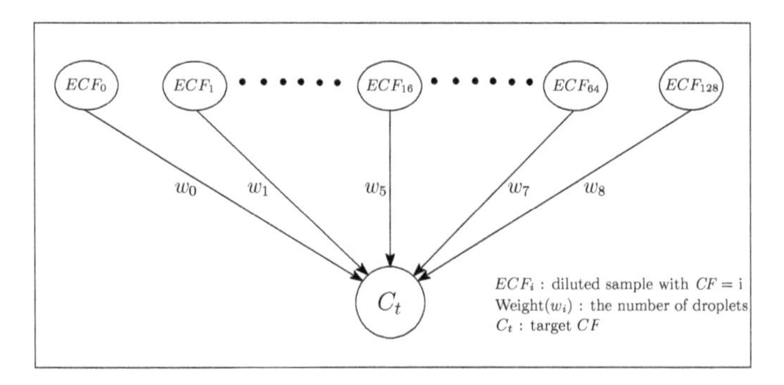

Figure 8.6: The ILP-model (with permission from ACM [119])

Let us assume that w_i ($i = 0, 1, \cdots, n$, where n is the accuracy level) denotes the total number of aliquots of size $\frac{1}{16}$X required for each ECF (Fig. 8.6). We define w_i as free variable ($w_i \in \mathbb{Z}_{\geq 0}$), i.e., for realizing the desired target-*CF*, the ILP-solver assigns certain values to w_i and the objective function becomes:

$$Minimize : w_0 + w_1 + w_2 + \cdots + w_n \tag{8.1}$$

Let c_i denote the exponential concentration factors (ECFs) for $i = 0, 1, \cdots, n$. Note that for a particular target-*CF* there exist several combinations of ECFs that can be used to generate it. For example, we can generate $CF = \frac{6}{8}$ by selecting a combination from $\{(\frac{4}{8}, \frac{8}{8}), (\frac{2}{8}, \frac{8}{8}, \frac{8}{8}), (\frac{4}{8}, \frac{4}{8}, \frac{8}{8}, \frac{8}{8}), (\frac{2}{8}, \frac{4}{8}, \frac{8}{8}, \frac{8}{8}, \frac{8}{8}), \cdots\}$. However, if we choose the $(\frac{4}{8}, \frac{8}{8})$ pair, the cost is minimized. So, the total number of aliquots

(k) needed for the target-droplet becomes 2, i.e., the volume of the target-droplet becomes $\frac{2}{16}$X, where $k \in \mathbb{Z}_{\geq 1}$. Apart from cost minimization, we require that the generated target-CF should not exceed the error-tolerance limit $(E_{tr}) = \frac{1}{2^{n+1}}$, i.e., we have:

$$\sum_{i=0}^{n} c_i \times w_i - C_t \times k < E_{tr} \times k \tag{8.2}$$

$$\sum_{i=0}^{n} c_i \times w_i - C_t \times k > -E_{tr} \times k \tag{8.3}$$

$$\sum_{i=0}^{n} w_i - k = 0 \tag{8.4}$$

Example 19 If we run the ILP code for producing target-CF = $\frac{71}{128}$, the required ECFs become $\{\frac{2}{128}, \frac{8}{128}, \frac{32}{128}, \frac{128}{128}\}$ and the corresponding number of aliquots are 1, 1, 1, 3, respectively. So, the total number of aliquots becomes 6, i.e., the volume of the target-droplet becomes $\frac{6}{16}$X. Thus, the ILP-based method consumes $\frac{3.33}{16}$X-volume sample, $\frac{2.67}{16}$X-volume buffer, produces zero-waste, and requires 6-units of mixing-time. Thus, it drastically reduces the cost of sample-preparation compared to the "split-less" TWM method discussed earlier.

Note that the computational overhead of the ILP method may increase rapidly with the increase of n. In order to cope with it, a fast heuristic method can be used to solve this problem. In order to facilitate the process, we use the concept of skewed mixing-tree introduced in the REMIA scheme [60]. Generally, REMIA works in two phases. In the first phase, REMIA creates a skewed mixing-tree for the target-CF (for reactant minimization), and in the next phase, it creates a dilution-forest for providing the corresponding ECFs as mandated by the leaf nodes of the skewed mixing-tree. An example for producing target-CF = $\frac{71}{128}$ using the REMIA scheme [60] is shown in Fig. 8.7.

In the heuristic method, a dilution tree is constructed as follows. Initially, it creates a skewed mixing tree using REMIA, and then, it converts the skewed mixing tree into a split-less mixing tree by assigning appropriate input demands to the former. The split-less scheme on REMIA with MEDA biochips, called SL_REMIA, may increase the volume of the target-droplet compared to the ILP-method. However, it reduces the size of the target droplet significantly compared to split-less TWM (Fig. 8.8). In order to make the comparison meaningful, we assume that the minimum volume aliquot-droplet supported on MEDA is $\frac{1}{16}$X. We will show later that this scheme can be easily parallelized for the production of multiple-CFs. Adapting a split-less scheme without incurring exponential reactant-cost is not possible on conventional DMFB-platforms since they are not suitable for manipulating droplets with fractional volumes.

Example 20 An example for generating target-CF =$\frac{71}{128}$ using the heuristic method is shown in Fig. 8.9. For supplying the input-droplets of the dilution tree, droplets

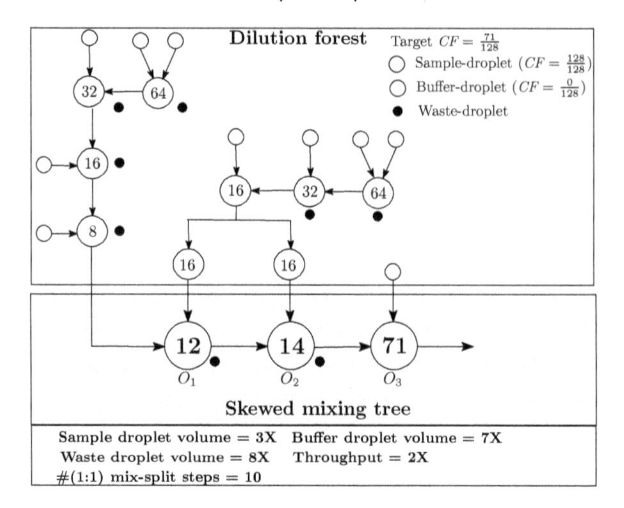

Figure 8.7: Generating target-CF = $\frac{71}{128}$ for accuracy level 7 using traditional REMIA [60] on a DMFB platform (with permission from ACM [119])

with ECF-values $\frac{16}{128}$ and $\frac{8}{128}$ can be generated in advance, using only the channel dispensing operation supported by MEDA-chips. In the top-part of the figure, we show how droplets with ECF-values $\frac{16}{128}$ and $\frac{8}{128}$ are produced without using any split operation. We create a complete dilution tree for the respective ECFs, and store them in reservoirs from which a dispensing operation is used to emit the required volume

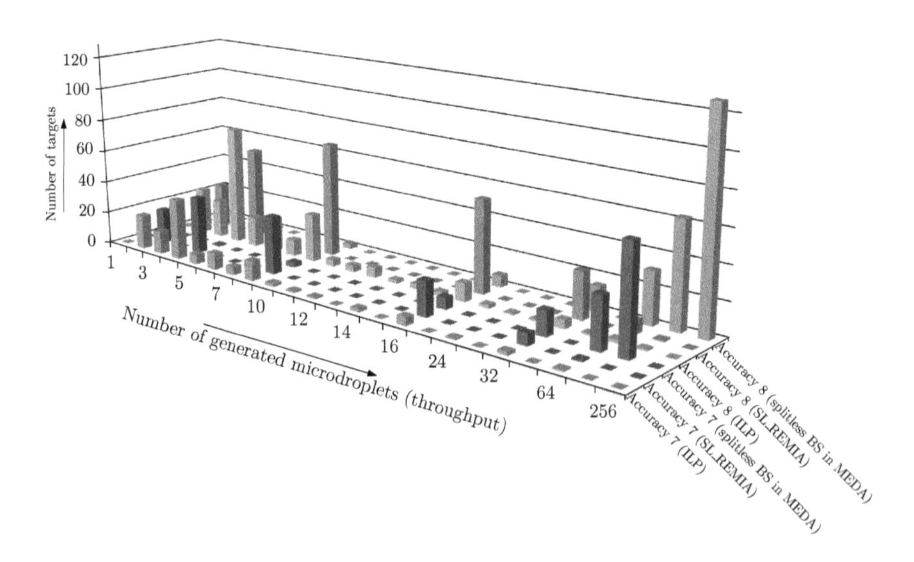

Figure 8.8: Volume of target-droplets obtained by split-less TWM, SL_REMIA, and ILP-method on MEDA biochips (with permission from ACM [119])

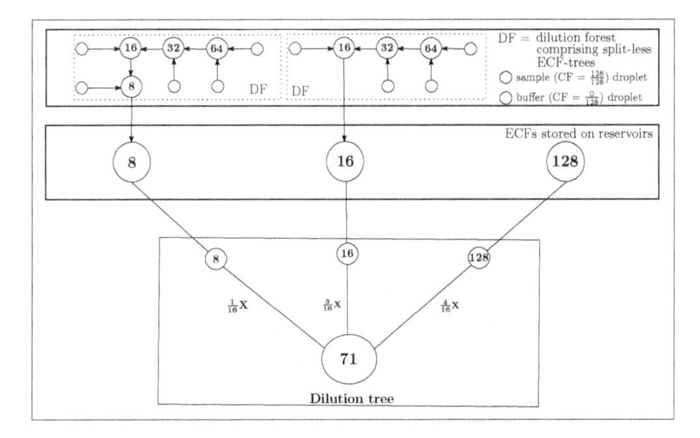

Figure 8.9: Generation of target-$CF = \frac{71}{128}$ with accuracy level 7 on MEDA by MTM (with permission from ACM [119])

droplets. Next, the dilution tree (shown in the bottom-part of Fig. 8.9) is executed for generating the target-droplet. This process eventually obviates all split-operations and produces target-$CF = \frac{71}{128}$ requiring $\frac{4.48}{16}$X samples, $\frac{3.52}{16}$X buffers, zero waste, and 7-units of mixing time[7] – thus solving the task robustly with moderate cost. Besides, this approach additionally allows to produce multiple target-*CF*s robustly on the same framework. By virtue of the inherent characteristics of the REMIA-algorithm, multiple target-*CF*s would require only different ECFs as inputs. Thus, droplets with all ECFs are produced in advance and stored in reservoirs. We can easily plug them together on MEDA, in parallel, to create dilution trees for desired target-*CF*s. Note that no split-operation is needed, and hence, no waste-droplets are produced. Unlike traditional multi-target sample-preparation approaches [100], there is no issue of sharing intermediate droplets among various mixing trees for cost reduction – which also eliminates the possibility of split-errors.

The proposed method reduces the dispensing time noticeably by reducing the number of dispensing steps compared to traditional REMIA. For example, REMIA performs four dispensing operations for feeding the inputs of the skewed mixing tree as shown in Fig. 8.7. The heuristic (ILP) method requires only three (four) dispensing operations for execution of the dilution tree (Figs. 8.6 and 8.9). This is due to the power of MEDA-specific dispensing operations that allow to dispense a certain volume of droplet with a particular *CF* to the chip.

8.3 EFFECT OF DISPENSING ERRORS ON TARGET-*CF*

Although MTM eliminates the possibility of split-errors, volumetric errors may still occur during dispensing operations. A small amount of dispensing error might affect the desired target-droplet badly (Fig. 8.3), even in the absence of mix-split errors.

[7]Mixing operations are performed between micro-volume MEDA droplets. Also, they can be performed in any order since no intermediate waste-droplets are produced.

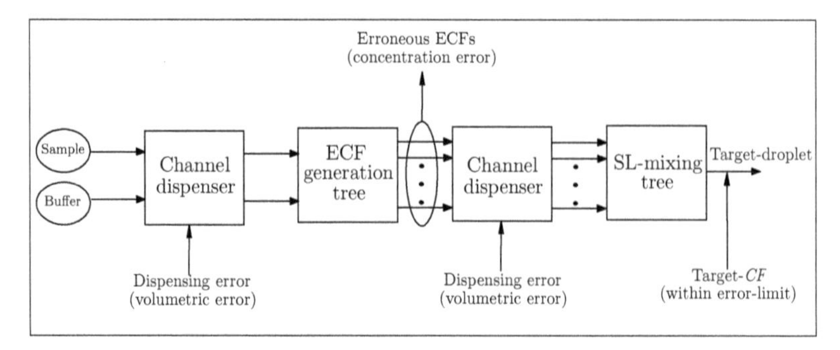

Figure 8.10: Overall flow of MTM (with permission from ACM [119])

Dispensing errors may occur in both phases, i.e., while creating the ECFs from sample and buffer droplets as governed by the dilution tree, and during the preparation of the desired target-*CF* in accordance to the split-less mixing tree (constructed by ILP/SL_REMIA) with ECF values as inputs. Thus, dispensing errors may occur in $3^{(d_1+d_2)}$ ways[8] where d_1 (d_2) denotes the number of dispensing operations in the first (second) phase. The overall flow of the MTM is shown in Fig. 8.10.

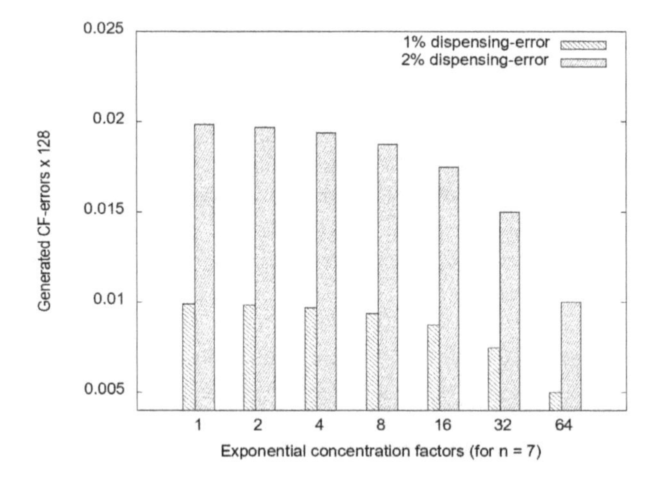

Figure 8.11: Maximum *CF*-errors among all combinations of dispensing errors for accuracy level 7 (with permission from ACM [119])

We now describe a method for tolerating the effect of dispensing errors on the target-droplet. Initially, we did experiments to ascertain the overall effect of various dispensing errors considering +, -, and ϕ (with 1% and 2% error). We assume that the demand of each individual ECFs is 8X for accuracy level 7. While doing experiments, we recorded the maximum *CF*-errors for all ECFs and reported the results in Fig. 8.11. It can be observed that the *CF*-error increases significantly when the

[8] considering all possible combinations of erroneous droplets, i.e., larger-volume (+) dispensed droplet, smaller-volume (-) dispensed droplet, and correct-volume (ϕ) dispensed droplet

magnitude of the dispensing-error increases to 2% from 1%. We further perform experiments to observe the effects of 1% and 2% dispensing-errors on the entire range of target-*CF*s for accuracy level = 7 (without applying any error-tolerance scheme). The concentration error is found to exceed the error-tolerance range for a large number of target-*CF*s (see Fig. 8.12).

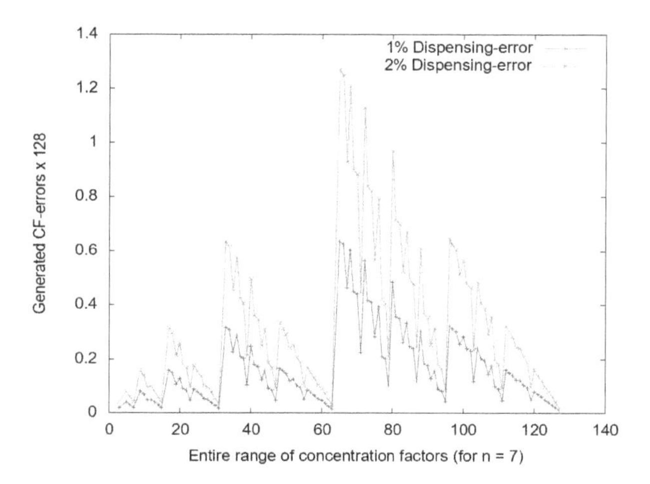

Figure 8.12: *CF*-errors (without applying any error-tolerance method) for all target-*CF*s with accuracy level 7 (with permission from ACM [119])

Note that MEDA-biochips allow to control the volume of the dispensed droplet precisely by counting the number of microelectrodes comprising the configured active zones. For example, one can dispense $\frac{1}{2}$X-volume droplet by configuring 2×4 microelectrodes (as shown in Fig. 8.1). It breaks the channel along the cut-point once the configured electrodes are filled up. Thus, this type of dispensing operation bears the characteristics of both continuous-flow and droplet-based system. Hence, while breaking the channel, the two droplets touching the cut-point may get affected by the volumetric-error, if any, during a dispensing operation. So, in the case of ε_d volumetric-error, the droplets appearing on the two sides of the cut will have volume $(1+\varepsilon_d)$ and $(1-\varepsilon_d)$, out of which, only a single-droplet is used $(\pm\varepsilon_d)$ in the subsequent operation. Therefore, while dispensing N droplets for an ECF, the volume of just a single droplet is affected. Before execution of the assay, we compute the *CF*-error in the target-droplet using Expression 8.5. If the *CF*-error exceeds the error-tolerance limit, it doubles the volume of the target-droplet until the newly calculated *CF*-error lies within the allowable error-tolerance limit. The volume of each ECF increases with the increase of the volume of a target-droplet. Alternatively, it means that the number of each input ECF is doubled when we double the volume of the target-droplet. Thus, a larger number of error-free ECF droplets participate in the dilution process. As a result, every time when we double the target-droplet volume, the *CF*-error in the target-droplet is reduced. The MTM method is simulated to check the set of affected target-*CF*s for accuracy 7 (Fig. 8.12). It is observed that it bounds the

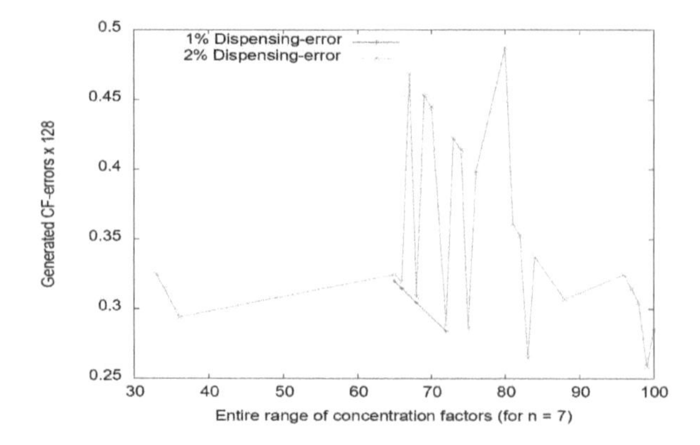

Figure 8.13: Maximum *CF*-error for all affected target-*CF*s of Fig. 8.12 (with permission from ACM [119])

error for all affected target-*CF*s (Fig. 8.13). The characteristics of MTM in contrast to TWM and REMIA are highlighted in Table 8.3.

Table 8.3
Comparative characteristics of MTM against prior art.

Method	Underlying architecture?	Errot-tolerant?	Waste-free?	Split-less?	Considers dispensing errors?	Reduces dispensing operations?	Target-droplet volume
TWM [163]	DMFB	No	No	No	No	No	2X
REMIA [60]	DMFB	No	No	No	No	No	2X
MTM [119]	MEDA	Yes	Yes	Yes	Yes	Yes	$\frac{2}{16}$X - $\frac{64}{16}$X*

* The volume of the target-droplet lies within 2X for 99% *CF*s for the accuracy 7. A detailed distribution is shown in Fig. 8.17 for accuracy 7, 8 and 9.

8.4 PROBLEM FORMULATION

The MTM procedure is now formally described below:

Input: 1) A supply of sample and buffer; 2) target-$CF = C_t$; 3) the dimension of the MEDA region that emulates the area of one electrode for a traditional DMFB; and 4) microfluidic module library representing type, size, and operation-time for performing fluidic functions on the chip.

Output: Dilution-forests (for ECFs) and skewed mixing tree (for target-CFs), devoid of any split-operation that are required for producing multiple droplets with the target-CF.

8.5 RESULTING METHODOLOGY

Based on the input information, MTM produces a target-CF on a MEDA chip as follows:

- It first creates a dilution-tree (skewed mixing tree) for the given target-CF using ILP (REMIA [60]).
- If the tree is created following REMIA, it is transformed into a weighted *"split-less" dilution-tree* suitable for implementation on MEDA, and the required ECFs are noted.
- It determines the volume of each individual ECF that needs to be dispensed. Afterwards, it calculates the maximum concentration error (C_{er}) in the target-droplet using Expression 8.5. If C_{er} exceeds the error-tolerance range ($\pm\frac{0.5}{2^n}$), it doubles the required volume of the target-droplet until the newly calculated C_{er} becomes less than the error-tolerance range.
- Finally, the droplets with the required ECFs are dispensed from the reservoirs to be used as inputs to the split-less dilution tree.

MTM is essentially a two-phase process. Note that dispensing error may occur in both the phases but MTM can efficiently tolerate the volumetric errors caused by such erroneous dispensing operations. More precisely, MTM works as follows:

8.5.1 SPLIT-LESS ECF DILUTION FOREST

In this phase, droplets with each ECF are created using sample and buffer droplets. The exponential-dilution tree is modified by feeding appropriate-size droplets as inputs so that no waste-droplet needs to be produced following a mixing operations, and thus it guarantees split-less execution of the dilution tree. We illustrate the procedure with the following example: We assume that the size of a (4×4) MEDA microelectrode array is equal to that of a single DMFB electrode; hence fractional-volume droplets such as $\frac{1}{16}$X, $\frac{2}{16}$X, …, $\frac{14}{16}$X, and $\frac{15}{16}$X can be produced apart from 1X-, 2X-, 3X-, 4X-, $\cdots$ volume droplets.

Example 21 Let us assume that we need to generate a droplet with ECF = $\frac{16}{128}$ with volume = 4X to fill a reservoir. We initially construct a dilution tree using the conventional (1:1) mix-split sequence using sample and buffer droplets to produce the

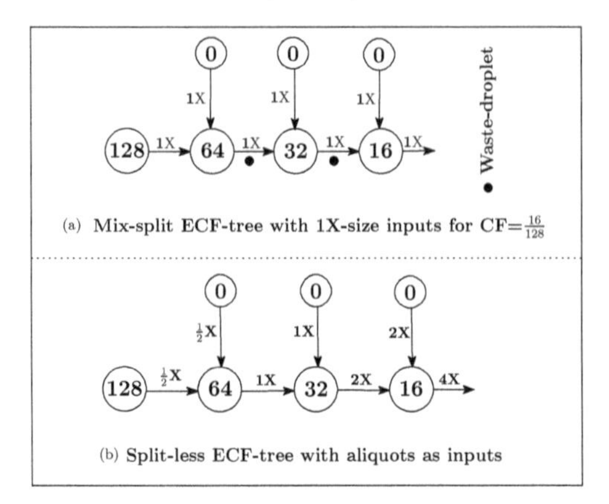

Figure 8.14: Generation of 4X-volume with ECF = $\frac{16}{128}$ for accuracy level 9 by MTM (with permission from ACM [119])

required ECF with the given accuracy level ($n = 7$) (see Fig. 8.14 (a)). This is based on an exponential dilution tree where the sample is mixed with buffer droplets iteratively with emission of 1X-size waste-droplet (shown as black dots) following a mix-split step. Note that the number inside each bubble represents the numerator of the corresponding *CF*; the denominator is implicitly 128, for $n = 7$. Next in Fig. 8.14(b), we show how a 4X-volume droplet with ECF = $\frac{16}{128}$ can be produced without using any split-operation by transforming the tree and feeding it with aliquots as inputs. The demand at the target node (with $CF = \frac{16}{128}$) is assigned as 4X. Next, it updates the demands of each remaining node with half of the demands of its parent node. At the completion of the process, the sizes of all aliquot-droplets become known. For example, we need $\frac{1}{16}$X, and $3 \times \frac{1}{16}$X- volume sample and buffer droplets, respectively, for producing 4X- volume of ECF = $\frac{16}{128}$. We present the outline of this procedure in Algorithm 8. Note that MTM can also generate different volumes of ECF = $\frac{16}{128}$ by changing the volumes of input droplets. For example, it can produce 1X-volume of ECF = $\frac{16}{128}$ by reducing the volume of each input droplet to $\frac{1}{4}$ (in Fig. 8.14 (b)). Also, it does not produce any intermediate waste-droplet due to the absence of split operations.

8.5.2 SPLIT-LESS DILUTION-TREE FOR THE TARGET-*CF*

In this phase, a MEDA-based split-less dilution-tree G is constructed using ILP or by modifying the corresponding DMFB-based mixing-tree. G is constructed in such a way that at least a 1X-volume target-droplet is produced. We can change this output-demand if multiple-droplets of the same *CF* are required in an application. Note that G requires only different-size aliquots of ECF-droplets, among which interpolated mixing operations will be performed. We discuss the working principle of this phase with an illustrative example below.

Algorithm 8: Split-less ECF dilution tree

Input: Desired exponential CF-value = E_{cf}, demand of $E_{cf} = d$, accuracy level = n, and micro-electrode sub-array-size = $N \times N$ (equivalent to one electrode of conventional DMFB)
Output: split-less *dilution-tree* for producing sufficient volume of $CF = E_{cf}$.

1 Let $O = \{O_1, O_2, O_3, \ldots, O_k\}$ be the nodes of a graph E_g whose nodes represent the CF values $\{2^n, 2^{n-1}, 2^{n-2}, \ldots, 2^1\}$;
2 Let $cf = \{cf_1, cf_2, cf_3, \ldots, cf_k\}$ be the unique CF values of E_g; //*where $cf[i] > cf[i+1]$*//
3 Remove all non-leaf nodes of E_g which have lower CF-value than E_{cf};
4 **if** $2^{len(E_g)-1} < d \times (N \times N)$ **then**
5   //*Update demand d of E_{cf} when input droplets volume in O_1 become $< \frac{1}{N \times N}$*//
6     $d = 2^{len(E_g)-1}$
7 $N_d(O[k]) = d$; //*$N_d(j)$: droplet volume demand in node j*//
8 Assign demand zero to the remaining nodes;
9 **for** $(i = |O|-1; i > 0; i = i-1)$ **do**
10     $E_d(l_c(O[i]), O[i]) = \frac{N_d(O[i])}{2}$;
11     $E_d(r_c(O[i]), O[i]) = \frac{N_d(O[i])}{2}$; //*$E_d(j,k)$: input droplet demand from node j to node k; $l_c(j)$: left child of j; $r_c(j)$: right child of j *// $N_d(O[i-1]) = \frac{N_d(O[i])}{2}$;
12 **return** E_g

Example 22 Let us continue with the example target-$CF = \frac{71}{128}$ for accuracy level $n = 7$. MTM first creates a dilution-tree using ILP or a skewed mixing tree using *REMIA* [60] for the desired target-CF. It then converts the skewed mixing tree (if obtained by REMIA) into a MEDA-based split-less mixing-tree G as shown in the bottom section of Fig. 8.9; otherwise it uses the dilution-tree obtained by the ILP-method. We assign the output-demand as 1X, and continue computing the input-demand backwards to determine the respective size of each input ECF-aliquot. It has been observed that it requires ECFs = $\{\frac{8}{128}, \frac{16}{128}, \frac{128}{128}\}$ with $\{\frac{2}{16}$X, $\frac{6}{16}$X, $\frac{8}{16}$X$\}$- volume aliquots for producing 1X-volume target-droplet. The overall flow for constructing the MEDA-based *split-less mixing-tree* is shown in Algorithm 9.

8.6 ERROR-TOLERANCE

We now discuss the error-tolerance mechanism with the help of example $CF = \frac{71}{128}$. The aliquots with ECF = $\{\frac{8}{128}, \frac{16}{128}, \frac{128}{128}\}$ need to be supplied from input reservoirs (Fig. 8.9), where the volume of the aliquots are $\{\frac{2}{16}$X, $\frac{6}{16}$X, $\frac{8}{16}$X$\}$, respectively. Alternatively, the volumes of ECFs = $\{\frac{8}{128}, \frac{16}{128}, \frac{128}{128}\}$ are equivalent to $\{2,6,8\}$ aliquots where the volume of an individual aliquot is $\frac{1}{16}$X. MEDA can efficiently supply the required volume of droplets using channel-dispensing operation [176]. However, erroneous dispensing operations may lead to some concentration error in the target-droplet. We use Expression 8.5 for calculating the maximum concentration error in the target-droplet:

$$C_t' = \frac{C_1(N_1\ \theta_1\ \varepsilon_d) + C_2(N_2\ \theta_2\ \varepsilon_d) + \cdots + C_i(N_i\ \theta_i\ \varepsilon_d)}{(N_1\ \theta_1\ \varepsilon_d) + (N_2\ \theta_2\ \varepsilon_d) + \cdots + (N_i\ \theta_i\ \varepsilon_d)} \tag{8.5}$$

where C_i denotes the concentration value of an ECF, N_i denotes the total number of aliquots of size $\frac{1}{16}$X required for C_i, ε_d denotes the magnitude of the volumetric

Algorithm 9: Split-less dilution-tree suitable for MEDA

Input: Desired target-$CF = C_t$, accuracy level n and micro-electrode sub-array size = $N \times N$ (equivalent to one electrode of conventional DMFB)
Output: *split-less mixing tree* (G) for the target-$CF = C_t$

```
1  Express C_t as x/2^n;
2  Let G represent the skewed mixing tree generated by REMIA [60] for the target-CF = C_t;
3  Let O = {O_1, O_2, O_3, ..., O_k} be the nodes of G, which represent the sequence of (1:1) mix-split steps;
4  for (i = 0; i < |O|; i = i+1) do
5      //Assignment of demands to nodes and edges of G;
6      N_d(O[i]) = (2^{i+1}/(N×N))X; E_d(l_c(O[i]),O[i]) = ⌈N_d(O[i])/2⌉; E_d(r_c(O[i]),O[i]) = ⌈N_d(O[i])/2⌉;
7  if N_d(O[k]) % (N×N) != 0 then
8      //If assigned volume of the target-droplet is less than 1X;
9      N_d(O[k]) = (2^N/(N×N))X;
10     E_d(l_c(O[k]),O[k]) = ⌈N_d(O[k])/2⌉;
11     E_d(r_c(O[k]),O[k]) = ⌈N_d(O[k])/2⌉;
12     for (i = |O| − 2; i > −1; i = i − 1) do
13         //Update the demands of other nodes and edges of G;
14         N_d(O[i]) = ⌈N_d(O[i+1])/2⌉;
15         E_d(l_c(O[i]),O[i]) = ⌈N_d(O[i])/2⌉;
16         E_d(r_c(O[i]),O[i]) = ⌈N_d(O[i])/2⌉;
17 return G
```

error ($\leq \pm 1\%$) [181, 193]) during dispensing operations, and θ_i denotes the use of the larger (+) or smaller (-) volume erroneous droplet.

Before the execution of the split-less dilution tree, MTM computes the concentration factor (C_t') of the target-droplet using Expression 8.5. Next, it calculates the maximum concentration error $C_{er} = |C_t - C_t'|$ in the presence of dispensing error, if any. For target-$CF = \frac{65}{128}$, initially the maximum CF-error becomes $\frac{0.64}{128}$ for $\frac{2}{16}$X-volume target-droplet which exceeds the allowable error-limit $\frac{0.5}{128}$. Hence, we consider doubling the volume of the target-droplet ($\frac{4}{16}$X) and re-calculate the concentration error. Eventually, the loop terminates since the concentration error becomes $\frac{0.32}{128}$, which is less than the allowable error-limit. Finally, we calculate the volume of each individual ECF that needs to be dispensed. In this example, the volumes of the ECFs = $\{\frac{2}{128}, \frac{128}{128}\}$ are $\{\frac{2}{16}$X, $\frac{2}{16}$X$\}$, respectively.

8.7 ERROR-FREE MULTIPLE TARGET SAMPLE PREPARATION

MTM is also capable of producing reliable multiple target-*CF*s on a MEDA platform. In order to do this, we first create a dilution-tree for each individual target-*CF*. Since, no waste-droplets are produced, all such trees can be executed on the MEDA-chip, in parallel, with no need for sharing intermediate droplets among different target-*CF*s. As MEDA comprises several hundreds of microelectrodes, enough area will be available for implementing them concurrently on-chip. Aliquots of required volumes are extracted from the reservoirs, where fluids with required ECFs were loaded *a-priori*. MTM can achieve higher throughput and supply a number of droplets for each target-*CF*, if required. Thus, it is not only capable of producing multiple

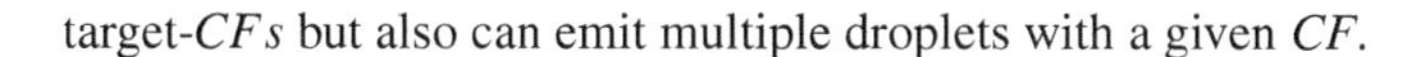

target-CFs but also can emit multiple droplets with a given CF.

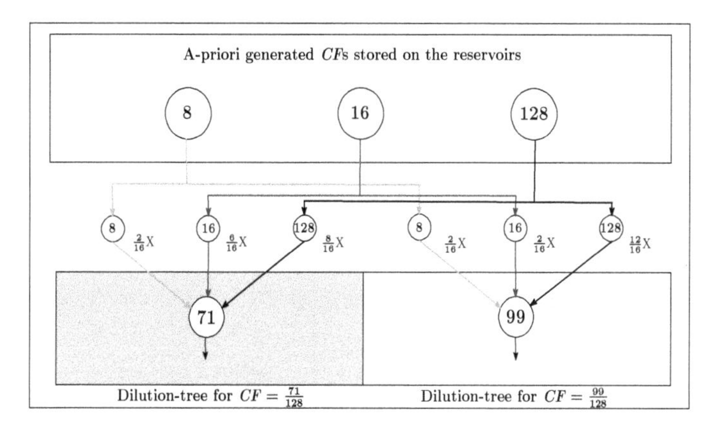

Figure 8.15: Robust dilution scheme for generating target-CFs = $\{\frac{71}{128}, \frac{99}{128}\}$ by MTM (with permission from ACM [119])

Example 23 Let us consider the production of target-CFs = $\frac{71}{128}$ and $\frac{99}{128}$ for accuracy 7. Fig. 8.15 demonstrates the overall flow. It shows the *dilution-tree* for each individual target-CF. Fluids with necessary ECFs are stored ahead of time on the reservoirs by executing the corresponding split-less ECF dilution forest. After that, the desired target-CFs are prepared in parallel by dispensing the required volume of aliquots from the reservoirs.

MTM thus efficiently utilizes ECF-droplets stored in the reservoirs. Note that their consumption will increase with the number of target-CFs or their output-demand. For example, larger volumes of target-droplets may be required for (i) screening a large number of patients in a point-of-care diagnostic environment, (ii) performing repeated execution of a bioassay (to enhance reliability of test results), and (iii) for conducting various bioanalytical tests [133]. Therefore, to achieve high throughput, it is desirable to store sufficient amount of ECF-droplets ($\geq 8X$) in the reservoirs to average out the overall cost of reactant-droplets. Since ECFs are used to generate multiple target-CFs in the second phase, from the collective perspective of multiple target production, any left-over ECFs are not truly wasted; they can be used to generate other target-CFs later on.

8.8 EXPERIMENTAL RESULTS

We have implemented MTM in Python and used a desktop computer with a 2 GHz Intel Core i5 processor, 8 GB memory, and 64-bit Ubuntu 14.04 operating system to perform all experiments. We run experiments for evaluating the effectiveness of MTM implemented on a MEDA-chip and compare results with TWM [163],

DMRW [131] and REMIA [60], CoDOS [99], WSPM [92], and MRCM [98] for conventional DMFB-platforms. We have used the IBM ILOG CPLEX [64] optimization tool for solving the ILP problem.

Table 8.4
Performance of MTM for generating ECFs for accuracy level $n = 9$.

Target-ECF	$CF_{1\%}$	$CF_{2\%}$	n_s	n_b	t_{dv}
$\frac{2}{512}$	$\frac{2.02}{512}$	$\frac{2.04}{512}$	$\frac{0.06}{16}$X	$\frac{15.94}{16}$X	1X
$\frac{4}{512}$	$\frac{4.02}{512}$	$\frac{4.04}{512}$	$\frac{0.13}{16}$X	$\frac{15.87}{16}$X	1X
$\frac{8}{512}$	$\frac{8.02}{512}$	$\frac{8.04}{512}$	$\frac{0.25}{16}$X	$\frac{15.75}{16}$X	1X
$\frac{16}{512}$	$\frac{16.02}{512}$	$\frac{16.04}{512}$	$\frac{0.50}{16}$X	$\frac{15.50}{16}$X	1X
$\frac{32}{512}$	$\frac{32.02}{512}$	$\frac{32.04}{512}$	$\frac{1}{16}$X	$\frac{15}{16}$X	1X
$\frac{64}{512}$	$\frac{64.02}{512}$	$\frac{64.04}{512}$	$\frac{2}{16}$X	$\frac{14}{16}$X	1X
$\frac{128}{512}$	$\frac{128.02}{512}$	$\frac{128.03}{512}$	$\frac{4}{16}$X	$\frac{12}{16}$X	1X
$\frac{256}{512}$	$\frac{256.01}{512}$	$\frac{256.02}{512}$	$\frac{8}{16}$X	$\frac{8}{16}$X	1X

$CF_{1\%}$: ECF in the presence of 1% dispensing-errors;
$CF_{2\%}$: ECF in the presence of 2% dispensing-errors;
$n_s(n_b)$: the number of sample (buffer) droplets; t_{dv}: target-droplet volume.

We assume that the size of the 4×4 MEDA micro-electrode array is same as the size of one conventional DMFB electrode. Therefore, the minimum allowable droplet-volume is $\frac{1}{16}$X. Initially, we have evaluated the performance of MTM for generating the ECFs and report the average cost (per 1X-volume target-droplet) in Table 8.4. We also performed experiments to observe the effect of possible dispensing-errors on the ECFs. During experiments we have considered all values of ECFs and reported the maximum erroneous concentration value for each ECF in Table 8.4, when 1% and 2% volumetric-errors are injected during dispensing operations.

8.8.1 SINGLE-TARGET SAMPLE PREPARATION

We performed experiments on various real-life and synthetic test cases for evaluating the performance of MTM. We assume that a (1:1) mixing operation considering $\frac{1}{16}$X (1X) micro (macro)-volume MEDA (DMFB) droplet requires 1 (16) unit(s) of time. We have reported the number of consumed sample droplets (n_s), buffer droplets (n_b), mixing time (m_t), target-droplet volume (t_{dv}) for several concentration factors (CF') in Table 8.5 for real-life and synthetic test-cases with dispensing error (2%) injected in both the phases. In addition, we have taken erroneous ECFs (having maximum CF-error) for generating the target-CFs and reported the maximum CF-error. We observe that MTM efficiently generates the target-CFs within the allowable

Table 8.5
Performance of MTM.

Real-life test-cases						
CF	CF'	m_t	n_s	n_b	t_{dv}	$\|CF\text{-}CF'\| < \frac{0.5}{512}$?
$\frac{51}{512}$	$\frac{51.26}{512}$	7	$\frac{0.8}{16}$X	$\frac{7.2}{16}$X	$\frac{8}{16}X$	Yes
$\frac{358}{512}$	$\frac{358.26}{512}$	60	$\frac{44.75}{16}$X	$\frac{19.25}{16}$X	$\frac{64}{16}X$	Yes
$\frac{456}{512}$	$\frac{456.28}{512}$	28	$\frac{28.5}{16}$X	$\frac{3.5}{16}$X	$\frac{32}{16}X$	Yes
Synthetic test-cases						
CF	CF'	m_t	n_s	n_b	t_{dv}	$\|CF\text{-}CF'\| < \frac{0.5}{512}$?
$\frac{13}{512}$	$\frac{13.10}{512}$	03	$\frac{0.10}{16}$X	$\frac{3.9}{16}$X	$\frac{4}{16}X$	Yes
$\frac{75}{512}$	$\frac{75.31}{512}$	14	$\frac{2.34}{16}$X	$\frac{13.66}{16}$X	$\frac{16}{16}X$	Yes
$\frac{231}{512}$	$\frac{231.26}{512}$	31	$\frac{14.44}{16}$X	$\frac{17.56}{16}$X	$\frac{32}{16}X$	Yes
$\frac{391}{512}$	$\frac{391.28}{512}$	60	$\frac{48.88}{16}$X	$\frac{15.12}{16}$X	$\frac{64}{16}X$	Yes
$\frac{451}{512}$	$\frac{451.29}{512}$	60	$\frac{56.38}{16}$X	$\frac{7.62}{16}$X	$\frac{64}{16}X$	Yes

CF': target-droplet with maximum concentration error; m_t: mixing time; $n_s(n_b)$: number of sample (buffer) droplets; t_{dv}: target-droplet volume.

error-tolerance range without producing any waste-droplets. MTM does not require any sensing operation to verify the correctness of intermediate droplets when performed in an error-prone environment. This is due to the elimination of the sources of all volumetric split-errors following mixing operations.

We compare MTM with TWM [163], REMIA [60], CoDOS [99], WSPM [92], and MRCM [98]. In order to make the comparison fair, we assume that previous sample-preparation algorithms [60,92,98,99,163] are devoid of any volumetric split-errors and dispensing errors.

We also run simulation experiments over the entire range of 511 target-*CFs* for accuracy level = 9 and noted the consumption of reactant, the number of mixing operations, and waste-droplets. Simulated results are reported in Fig. 8.16. It can be seen that MTM outperforms prior methods in terms of these parameters. We also report the consumption of sample and buffer droplets, and mixing times over the entire range of target-*CFs* with accuracy 7, 8, and 9, in Table 8.6 (considering 2% dispensing errors in both phases).

8.8.2 MULTI-TARGET SAMPLE PREPARATION

In order to evaluate the effectiveness of MTM for producing multiple dilutions, we run experiments with several real-life and synthetic multiple target-*CFs* method and computed the relevant parameters as before. We have studied dilution gradients such

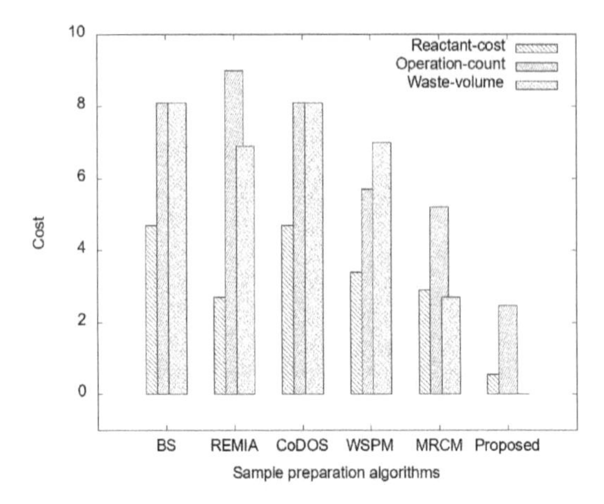

Figure 8.16: Comparison of MTM with previous sample-preparation algorithms (with permission from ACM [119])

Table 8.6
Performance of MTM for accuracy level n = 7, 8, and 9.

Accuracy	n_s	n_b	m_t
6	$\frac{4.16}{16}$X	$\frac{2.11}{16}$X	5.24
7	$\frac{6.57}{16}$X	$\frac{3.40}{16}$X	8.46
8	$\frac{11.69}{16}$X	$\frac{6.11}{16}$X	15.96
9	$\frac{23.13}{16}$X	$\frac{12.13}{16}$X	32.16

as linear, harmonic, logarithmic, Gaussian, and parabolic [72, 87], which are frequently used in titration

process [17]. Note that for linear dilution gradient, target-*CF*-values follow the pattern of arithmetic progression. For example, Bradford protein assay [17] requires *CF*s of (10%, 20%, 30%, 40%, 50%) of a diluted sample. Also, a linear dilution of *CFs* ranging from 10% to 40% is required in sucrose gradient analysis [17]. In *K*-fold exponential dilution, target-*CF*s assume the pattern as 1, $\frac{1}{K}$, $\frac{1}{K^2}$, $\cdots$. For example in qPCR [17], an eight-fold dilutions of yeast genomic DNA are required. For harmonic gradients, target-*CFs* become reciprocals of successive integers (1, $\frac{1}{2}$, $\frac{1}{3}$, $\cdots$). We approximated target-*CF* values for different dilution gradients in Table 8.7. For simplicity, we have shown only the numerator value of each target-*CF* in Table 8.7, where the denominator is implicitly assumed as 512. For synthetic test-cases, we randomly generate target-sets of different sizes (5, 10, 20, 30, 40, 50, 100) and report

Table 8.7
Target-*CF*s for different dilution gradients for accuracy = 9.

Dilution gradient system	Target-*CF*s
qPCR[a]	{256,32,4}
Bradford protein assay[b]	{51,102,154,205,256}
Sucrose gradient analysis[c]	{51,77,102,128,154,179,205}
Harmonic[d]	{256,171,128,102,85,73,64,57}

Table 8.8
Performance of MTM for multi-target real-life test-cases for accuracy = 9.

Dilution gradient	n_s	n_b	m_t
a	$\frac{3}{16}$X	$\frac{159}{16}$X	143
b	$\frac{21.01}{16}$X	$\frac{21.01}{16}$X	59
c	$\frac{33.40}{16}$X	$\frac{74.59}{16}$X	98
d	$\frac{18.96}{16}$X	$\frac{58.03}{16}$X	68

*Mixing is performed between micro-droplets of size $\frac{1}{16}$X;

the corresponding results in Tables 8.8 and 8.9 for real-life and synthetic test-cases for accuracy level 9.

Finally, we conduct an experiment for determining the volume of target-droplets to be produced for achieving the error-tolerance limit. We carry out experiments for the entire range of *CF*s for accuracy level 7, 8 and 9 and report the results in Fig. 8.17. It has been observed that 93% (74%) target-*CFs* with accuracy 7 (8) require 1X-volume DMFB-droplet to achieve the allowable error-tolerance range. For 2X-volume DMFB-droplets, 71% target-*CF*s achieve error-tolerance for accuracy 9.

Table 8.9
Performance of MTM for multi-target synthetic test-cases for accuracy level $n = 9$.

#Target-*CFs*	n_s	n_b	m_t
5	$\frac{79.92}{16}$X	$\frac{56.07}{16}$X	131
10	$\frac{238.68}{16}$X	$\frac{117.32}{16}$X	343
20	$\frac{1404.55}{16}$X	$\frac{527.44}{16}$X	1886
30	$\frac{743.23}{16}$X	$\frac{442.43}{16}$X	1140
40	$\frac{1006.38}{16}$X	$\frac{289.46}{16}$X	1436

*Mixing is performed between micro-droplets of size $\frac{1}{16}$X;

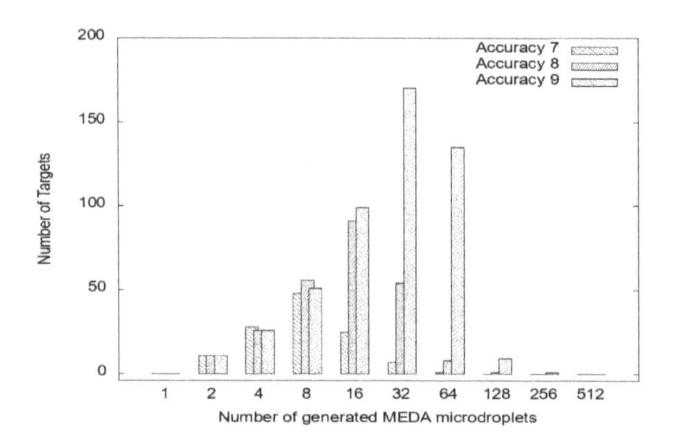

Figure 8.17: Volumes of target-droplets required to achieve error-tolerance by MTM for accuracy level n = 7, 8 and 9 (with permission from ACM [119])

8.9 CONCLUSIONS

Most of the errors in sample preparation are attributed to volume variation in droplets, which are caused by mix-split operations resulting in a change in the desired target concentration factor of the target solution. Volumetric imbalance among daughter-droplets is inherent to a "split" operation, which is an essential primitive in droplet-based microfluidics. Such errors are caused by random events and cannot be rectified so easily on a DMFB-platform. Traditional recovery processes include invoking of either "rollback" or "roll-forward" action when such errors are sensed online. Both of them increase reactant-cost or assay-completion time. MTM is based on a fundamentally different approach to robust sample preparation that eliminates the need for any droplet-splitting operation. Instead of using conventional mix-split sequences, we recast sample-preparation algorithms from a new perspective that rely on using only "aliquots-and-mix", and utilize the granularity of MEDA-chips to implement the procedure. As a by-product of this method, multiple target-CFs can also be produced in parallel on a MEDA chip. MTM does not produce any waste-droplet of intermediate concentration factors. Experimental results on various test-cases indicate that it reduces reactant-cost and mixing time significantly compared to previous approaches, while warranting the robustness of the process even when dispensing errors occur. As a future research problem, it will be interesting to investigate how such a "split-less" approach can be extended to handle mixture preparation with three or more fluids in the presence of various errors.

Section IV

Summary

9 Summary and Future Directions

In this book we have introduced the basics of algorithmic sample preparation using digital microfluidic technology and discussed various design issues that need to addressed for enabling the automation of biochemical protocols on LoCs. During the execution of an assay on-chip, different types of fluidic errors may occur compromising the reliability of operations, and they in turn, may jeopardize the outcome of the assay. We have analyzed in detail, the impact of volumetric split-errors on reaction-paths of assays, and how they corrupt the desired concentration factor of a target sample. Such an analysis not only provides deep insight on how the errors affect the accuracy of sample preparation but also leads to novel algorithmic solutions to error recovery on a digital microfluidic biochip.

Earlier approaches to error-correction in cyber-physical microfluidic systems are essentially based on the principle of "rollback", i.e., on sensing the presence of an error, attempts are made to re-execute a portion of the assay from a previous checkpoint. Unfortunately, there are uncertainties on the required number of rollback-attempts, and inadequate estimate of how much extra storage might be needed to store back-up droplets, particularly, in the presence of multiple fluidic errors. Thus, the rollback strategy for error management is non-deterministic with respect to both time and space.

In Chapter 1, we have introduced the basic architecture of conventional digital microfluidic biochips (DMFB), and those based on Micro-Electrode Dot Architecture (MEDA). Next, in Chapter 2, we have reviewed various methods for algorithmic sample preparation with DMFBs and demonstrated the effects of volumetric split-errors on the accuracy of target concentrations. A summary of prior error-recovery techniques for digital microfluidic biochips and MEDA is presented in Chapter 3.

In Chapter 4, we have discussed a "roll-forward" approach for DMFBs that guarantees cancellation of all volumetric split-errors when target-droplets are produced [122]. Unlike previous approaches, no rollback to past checkpoints or back-up droplets are needed. This technique is referred to as Error-Correcting Sample Preparation (ECSP), and it can easily be implemented on a cyber-physical DMFB having sensor-driven feedback-control loops. It is assumed that imprecise droplet mix-split operations may cause volumetric imbalance between two daughter-droplets. In ECSP, the droplets produced following an erroneous mix-split step, are allowed to participate in the error-recovery process, instead of being discarded or re-mixed. An application-specific digital microfluidic biochip for supporting ECSP is also designed. Results on comparative performance evaluation are reported in this chapter.

In Chapter 5, a detailed analysis of multiple split-errors is presented [123]. We observe that for a single split-error, the smaller-size erroneous droplet impacts the

DOI: 10.1201/9781003219651-9

target-*CF* more dominantly compared to its larger-size sibling. However, it is difficult to ascertain which combination of multiple split-errors causes the maximum deviation in the target concentration without performing an exhaustive simulation.

In Chapter 6, we have presented a simple empirical rule for finding the worst-case multiple split-error that tends to maximize the error in a given target-concentration. Based on that, we have described a new approach called EOSP (Error-Oblivious Sample Preparation) that does not require any sensor to detect volumetric split-errors [124] while executing a dilution assay, yet warranting the desired accuracy of target concentrations.

In Chapter 7, we have discussed a method (MTD) for producing multiple target-*CF*s on-demand [121], without exceeding the error-tolerance range. In this scheme, a small subset of concentration factors is chosen *a-priori*, from which any other target-*CF* can be produced using only one or two mix-split steps. In order to minimize the size of the subset, an ILP-based method, an approximation algorithm, and a heuristic method have been described.

In Chapter 8, a multi-target dilution algorithm (MTM) for MEDA chips has been presented that does not need any rollback, roll-forward, or droplet-split operation [119]. Instead of performing traditional mix-and-split steps with integral-volume droplets, it executes only aliquoting-and-mix operations using differential-size aliquots. Since the method does not involve any split operation, sensing for volumetric error or any corrective action is not needed for robust sample preparation. Dispensing errors, if any, can also be handled with MTM.

In summary, this book will serve as a compendium of various methods that have been developed so far for error-tolerant sample preparation with digital microfluidic biochips. It includes most of the material from the very basics to recent advances reported in this emerging area. The work presented here may open up several future research directions. For example, efficient error-tolerant methods for mixture preparation are yet to be explored. Identification of critical errors in a mixing graph, in general, seems to be a complex problem. Error management in flow-based biochips during sample preparation also requires further investigation. The CAD-algorithms presented in this book will help in understanding sample preparation with digital microfluidic biochips and related issues concerning error tolerance.

Section V

Appendix

A Error-Correcting Sample Preparation with Cyber-physical DMFBs

A.1 CYBER-PHYSICAL SYSTEM

A cyber-physical digital microfluidic system consists of a biochip, a few on-chip sensors, a feedback mechanism, and an associated control software. Depending on the protocol to be executed on the chip and its physical design, an actuation sequence is generated and stored in the control memory. The sensors are capable for detecting or measuring certain physical or fluidic attributes of droplets, such as droplet volume, concentration, or the presence of a volumetric split-errors. Depending on the feedback received from the sensor, the control software invokes appropriate actuation sequences to correct the errors, if any, chooses the required reaction-path to complete the protocol.

Fig. A.1 shows the schematic of a cyber-physical digital microfluidic platform. The control software sends signals to the biochip to execute the assay; at the same time it receives feedback from on-chip sensors and these data are compared with predefined threshold values. When an error is sensed, the control software triggers necessary actions to change the reaction-path of the assay.

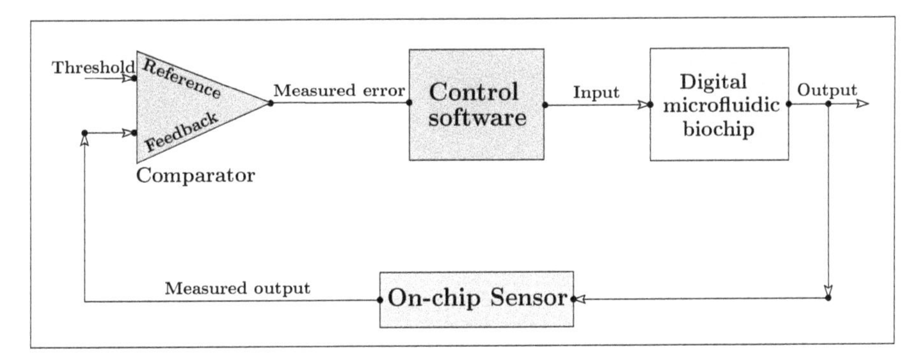

Figure A.1: Cyber-physical digital microfluidic biochip system (with permission from ACM [122])

DOI: 10.1201/9781003219651-A

A.1.1 SENSING SYSTEM

Many sensing systems can potentially be integrated on a digital microfluidic biochip for online monitoring of fluidic operations or attributes. A few of them are described below.

A.1.1.1 Optical Sensing

Optical sensing systems have been successfully used in the past [84,101,118,182]. In conventional laboratory environments, optical detection is preferred for robust and sensitive observations. Also, the availability of newer nanotechnology tools facilitates the integration of several micro-optical functions such as light source, lenses, wave-guides, and detectors on a microfluidic platform for detecting nanoscale analytes [84]. Details of a thin-film photo-detector integrated on a coplanar digital microfluidic biochip can be found in [77].

The outcome (erroneous/error-free) of a fluidic operation can also be evaluated by measuring droplet concentration through fluorescence techniques [38, 77, 105, 192]. Due to the attachment of fluorophore tag to a droplet, the wavelength of emitted light varies with droplet-concentration. An optical sensor consisting of light-emitting diode (LED) and photodiode is able to detect such changes [38, 151]. A thin-film InGaAs photo-detector-based digital microfluidic biochip can be found in [77].

Although optical detection remains an attractive technique for microfluidic applications, one of the challenging tasks is to integrate a low-cost sensitive optical detector on a microfluidic device. Also, when this type of detector is used for droplet sensing, some reagents are mixed with the droplet, which becomes unusable for later operations [4]. Furthermore, such a detector may be unable to locate the electrode position where the error has occurred.

A.1.1.2 Charge-Coupled Device (CCD)-based Sensing

CCD cameras can be conveniently integrated with microfluidic chips for capturing the top-view of droplets [105]. Also, droplets on multiple electrodes can be sensed simultaneously [117]. A schematic for CCD-based system for droplet sensing is shown in Fig. A.2 [105, 142].

The control software determines the location of droplets on-chip using a template matching technique. The image of a typical droplet is first used as a template; while searching for a match, the template image is moved across the entire 2D array-image to locate an equal-sized sub-image. After finding an equal-sized sub-image, a similarity measure called correlation index is then computed to find a match between the template and the selected sub-image as shown in Fig. A.3.

The images are stored in control memory in gray scale encoded as matrices or vectors. For example, if the template image has N pixels, then it can be represented as $\vec{X} = (x_1, x_2, \ldots, x_N)$, x_i is the gray level of a pixel. In similar fashion, we can represent the selected sub-image as $\vec{Y} = (y_1, y_2, \ldots, y_N)$; the correlation factor of

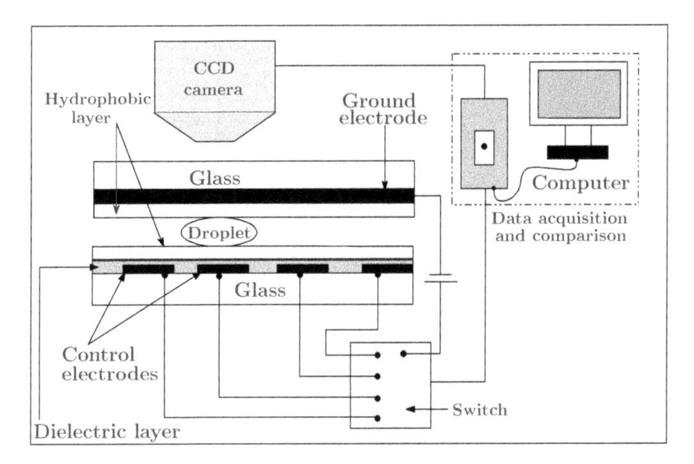

Figure A.2: A demonstration of CCD-based sensing system [105, 142] (with permission from ACM [122])

these two images can be calculated by the following expression [105].

$$C_f = \frac{\sum_{i=1}^{N}(x_i-\bar{x})\cdot(y_i-\bar{y})}{\sqrt{\sum_{i=1}^{N}(x_i-\bar{x})^2\cdot\sum_{i=1}^{N}(y_i-\bar{y})^2}} \tag{A.1}$$

In Expression A.1, $\bar{x}$ and $\bar{y}$ represent the average gray level value of the template and selected sub-image respectively. Higher value of C_f indicates a stronger relationship between the template and the selected sub-image, where $-1 \leq C_f \leq 1$. The correlation map of the template image and the original input image is obtained by finding the correlation factors for all positions in the image of the entire biochip. The droplet-locations can then be identified by searching for the largest correlation factors in the map [105]. Finally, the volume and concentration of droplets are estimated by analysing the CCD-images [59, 188].

CCD-camera-based sensing systems can accurately locate the positions of erroneous droplets and trigger the recovery action immediately as needed. However, in the case of photosensitive biochemical substances, e.g., fluorescent markers [130], CCD-camera-based sensors may not be suitable because the light source may affect the composition of the droplet.

A.1.1.3 Capacitive Sensing

In order to detect nanoliter-sized droplets efficiently, a capacitive sensing circuit can be integrated with a digital microfluidic lab-on-chip. A ring-oscillator-based capacitive sensing circuit was designed for detecting the presence/absence or the volume of a droplet [34, 59]. Fig. A.4 depicts a linear relationship between the capacitance measured by the ring oscillator and the droplet volume [46]. A supplementary video

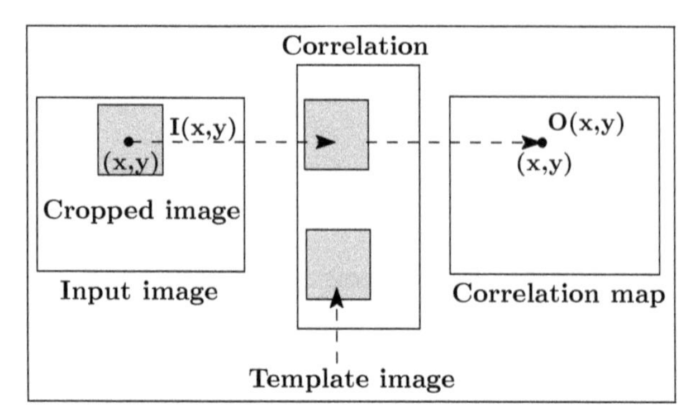

Figure A.3: An example of CCD-based sensing on a digital microfluidic biochip [105] (with permission from ACM [122])

on capacitance sensing in microfluidic systems [59] is available in the following *website*.

A specially designed capacitive sensing circuit to determine an error in droplet-volume or the volumetric difference of two nano-sized droplets is reported in [16]. In order to detect volumetric-error in a dispensed droplet, the sensed output voltage is compared with a reference voltage. If the difference exceeds some predefined threshold value (maximum allowable error), an error is signaled. For detecting a volumetric split-error, the output voltages from two capacitive-sensing circuits are fed to a differential comparator. In the roll-forward error-correcting technique discussed in Chapter 4, this differential sensing technique can be used to check whether or not a split operation is error-free. A block diagram of the sensing system is shown in Fig. A.5. Such a differential sensor is capable of detecting very small amount of split-error compared to other sensing schemes, and the sensing time is very small compared to droplet manipulation time.

A.1.2 INTEGRATION OF BIOCHIP AND CONTROL SOFTWARE

The overall cyber-physical system requires two interfaces, one for passing signals from sensors to the controller, and the other for transforming the $(0, 1)$ actuation matrix into a sequence of 2D voltage-map corresponding to the electrodes. While different kinds of sensors can be used to implement the first interface, a programmable logic controller (PLC) is, in general, used as the second interface, i.e., the one between the controller and the physical electrodes of the biochip. Sensors are placed at pre-located electrodes and droplet pathways are designed accordingly based on the designated checkpoints that are inserted on the reaction-path at different time instances. The PLC receives the actuation data from the controller memory, driven by sensor feedback, through the output ports of the controller [105].

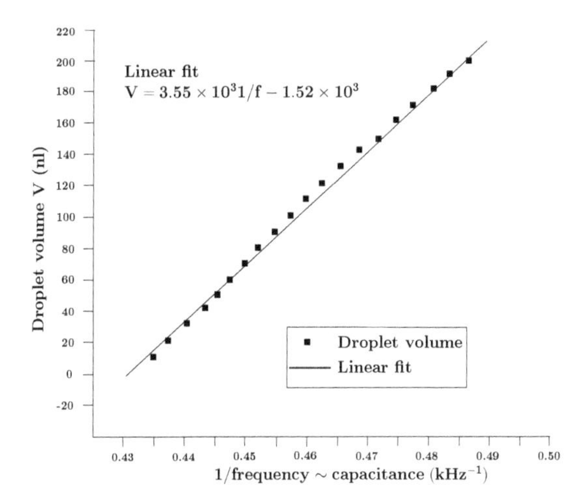

Figure A.4: Response of the ring-oscillator-based sensors in EWOD devices [46] (with permission from ACM [122])

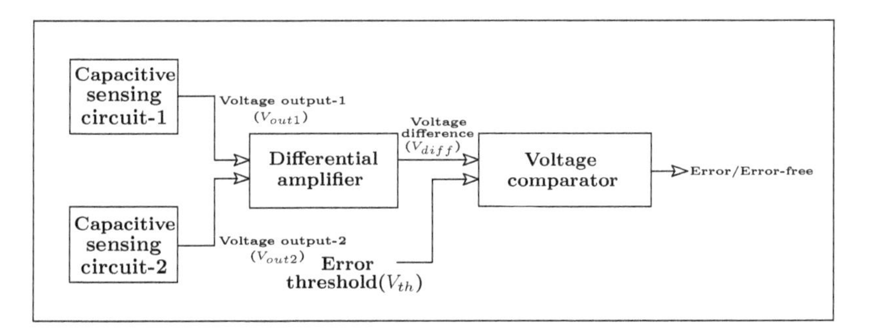

Figure A.5: Block diagram of the capacitive sensing circuit (with permission from ACM [122])

A.2 ERROR RECOVERY IN ROLLBACK AND ROLL-FORWARD APPROACHES

Consider the mix-split sequence (see Fig. 4.8, *Case I*) for target-$CF = \frac{241}{512}$ in the absence of any volumetric split-error. We describe below, the error-recovery sequence that will be invoked by the roll-forward strategy described in Chapter 4 and by a typical rollback approach.

Case I: Existing rollback approaches may allocate checkpoints to different stages of the mix-split sequence for possible errors detection, and re-execute the reaction-path from the previous checkpoint, when an error is sensed. However, for a sample-preparation assay, the concerned checkpoint where the error is detected, may not be critical, and hence, the corrective measures thus taken, are not at all necessary. Such issues of criticality have not been considered in rollback approaches. Hence, for this example, the rollback method may insert checkpoints at eight mix-split steps and rollback if needed. In contrast, the roll-forward technique allocates checkpoints only to a few intermediate stages, which are critical; in this case, for Steps 6, 7, and 8 only.

Case II: Consider now an occurrence of a single split-error (critical) on the reaction path. As shown in Fig. 4.8, *Case II*, rollback approaches will *re-do* the sequence of mix-split operations from the previous checkpoint for possible error-recovery. It is usually assumed that all errors will be corrected within a fixed number of such iterations. However, in reality, the error may still reappear, for example, in the last iteration. Furthermore, even for the best case (i.e., when the error is corrected in the first rollback-attempt), this approach requires more checkpoints as compared to ECSP. Moreover, rollback approaches need to store a large number of intermediate-droplets at different checkpoints on-chip, in order to facilitate error-recovery. For instance, if a split-error is sensed at Step S_7, the rollback method starts re-execution using the waste droplet that was produced at operation S_6 (assuming S_6 is the last checkpoint). On the other hand, the roll-forward method, on sensing an error, performs redundant operations $\{S'_8, S'_9\}$ along with $\{S_8, S_9\}$ with the erroneous droplets produced at Step S_7; hence erroneous droplets are utilized for canceling the error.

Case III: Now suppose two critical split-errors occur at steps $\{S_6, S_8\}$ as shown in Fig. 4.8 (*Case III*). In the rollback method, the error may be recovered by re-executing the reaction-path from ECSP $\{S_5, S_7\}$ using the waste droplets stored in the system. The roll-forward method first performs $\{S'_7, S'_8\}$ to compensate the effect of the first split-error that has occurred at Step S_6, and then performs Steps $\{S'_9, S''_9, S'''_9\}$ with the erroneous droplets produced at S_8, along with the original reaction-path, for canceling the errors. No intermediate-droplets need to be stored in this method; as soon as an error is sensed after a split-operation, the sibling-droplet is reused immediately along a parallel reaction-path.

Note that in rollback approaches, an attempt is made to correct the error at its source, whereas in the roll-forward approach, the error is corrected at the target. Furthermore, in the rollback (roll-forward) approach, the recovery method is

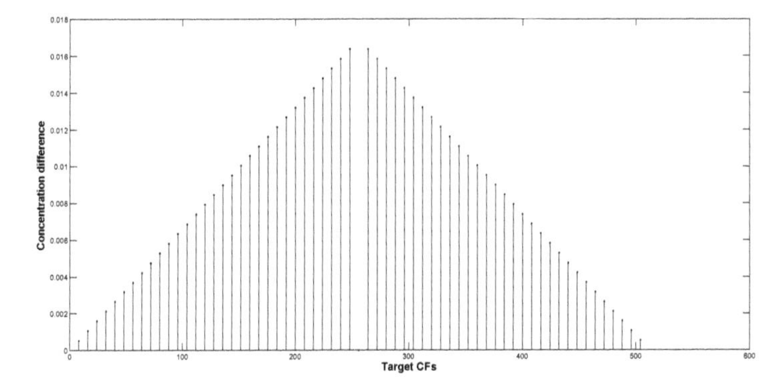

Figure A.6: Concentration-difference in the last-but-one step for some representative target-*CF*s with accuracy level $n = 9$ (with permission from ACM [122])

independent (dependent) on the target-*CF*. This follows from the fact that the corrective measures taken by the roll-forward method depend on the positions of critical mix-split steps on the reaction-path, which, in turn, depend on the target-*CF*.

A.3 SNAPSHOTS OF CONCENTRATION ERRORS

A volumetric split-error, which may occur in any step of the reaction path may lead to a concentration error, for a given target-*CF* C_t. It can be easily shown that the concentration error observed at the target-*CF* C_t due to an unbalanced split occurring at a particular step, is same as that for the target-*CF* $(1 - C_t)$. Fig. A.6 shows the symmetry property for some representative target-*CF*s when the accuracy level $n = 9$, under the volumetric split-error (7%) occurring at the last but one step. Also the locations of critical steps will be exactly the same for both targets C_t and $(1 - C_t)$.

A.4 SNAPSHOTS OF THE BIOCHIP LAYOUT AND SIMULATION

Fig. A.7 shows a (9×9) biochip layout on which our simulator was run. The reservoirs (inputs/outputs/waste collector) are placed around the boundary of the chip; there are four (3×3) mixer modules and sensors (red cells), which are placed in the interior as shown. While executing an assay, samples are first dispensed into the chip from the dispenser as shown in the first snapshot (first row of Fig. A.8); the next two snapshots show droplet-mixing on the mixer module. The first two snapshots of the second row depict the event of unbalanced-split. Next, two droplets after splitting are moved to the corresponding sensor locations for possible volumetric-error detection as shown in the last snapshot of Fig. A.8.

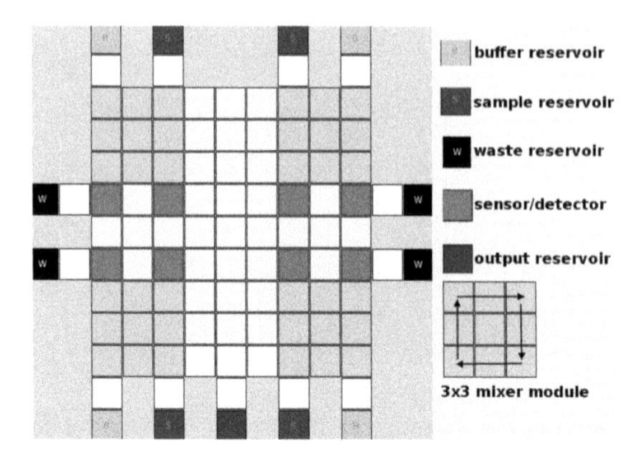

Figure A.7: An application-specific layout for the diluter chip (with permission from ACM [122])

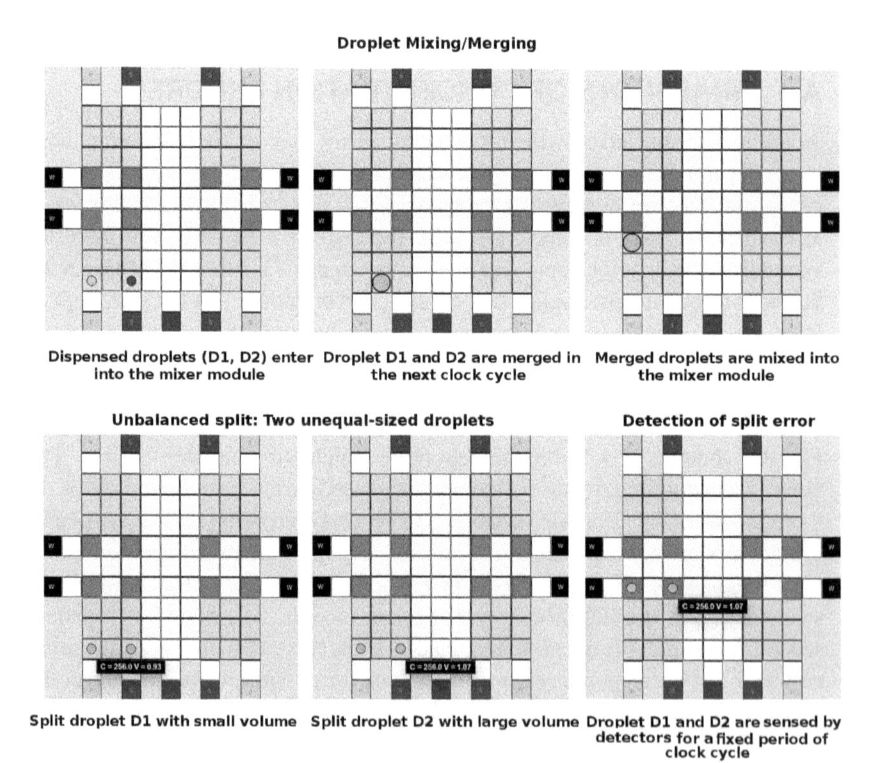

Figure A.8: Snapshots of on-chip split-error detection (with permission from ACM [122])

References

1. *The International Technology Roadmap for Semiconductors*, 2007 (accessed June 3, 2018). `http://www.itrs.net`.

2. abcam. 2014 (accessed June 3, 2018). `http://www.abcam.com/protocols/antibody-dilutions-and-titer`.

3. M. Alistar, E. Maftei, P. Pop, and J. Madsen. Synthesis of biochemical applications on digital microfluidic biochips with operation variability. In *Proc. of DTIP*, pages 350–357, 2010.

4. M. Alistar, P. Pop, and J. Madsen. Online synthesis for error recovery in digital microfluidic biochips with operation variability. In *Proc. of DTIP*, pages 53–58, 2012.

5. M. Alistar, P. Pop, and J. Madsen. Redundancy optimization for error recovery in digital microfluidic biochips. *Design Automation for Embedded Systems*, 19(1-2):129–159, 2015.

6. V. Ananthanarayanan and W. Thies. Biocoder: A programming language for standardizing and automating biology protocols. *Journal of Biological Engineering*, 4(1):13, 2010.

7. I. E. Araci and P. Brisk. Recent developments in microfluidic large scale integration. *Current Opinion in Biotechnology*, 25:60–68, 2014.

8. N. Azizipour, R. Avazpour, D. Rosenzweig, M. Sawan, and A. Ajji. Evolution of biochip technology: A review from lab-on-a-chip to organ-on-a-chip. *Micromachines*, 11(6), 2020.

9. T. Bell and B. McKenzie. Compression of sparse matrices by arithmetic coding. In *Proc. of DCC*, pages 23–32, 1998.

10. N. Bera, S. Majumder, and B. B. Bhattacharya. Analysis of concentration errors in sample dilution algorithms on a digital microfluidic biochip. In *Proc. of ICAA*, volume 8321, pages 89–100. Springer, 2014.

11. N. Bera, S. Majumder, and B. B. Bhattacharya. Simulation-based method for optimum microfluidic sample dilution using weighted mix-split of droplets. *IET-CDT*, 10(3):119–127, 2016.

12. S. Bhattacharjee, A. Banerjee, and B. B. Bhattacharya. Sample preparation with multiple dilutions on digital microfluidic biochips. *IET-CDT*, 8(1):49–58, 2014.

13. S. Bhattacharjee, A. Banerjee, T-Y. Ho, K. Chakrabarty, and B. B. Bhattacharya. On producing linear dilution gradient of a sample with a digital microfluidic biochip. In *Proc. of ISED*, pages 77–81, 2013.

14. S. Bhattacharjee, A. Banerjee, T-Y. Ho, K. Chakrabarty, and B. B. Bhattacharya. Efficient generation of dilution gradients with digital microfluidic biochips. *IEEE Trans. on CAD*, 2018.

15. S. Bhattacharjee, S. Poddar, S. Roy, J-D. Huang, and B. B. Bhattacharya. Dilution and mixing algorithms for flow-based microfluidic biochips. *IEEE Trans. on CAD*, 36(4):614–627, 2017.

16. B. B. Bhattacharya, S. Ghoshal, S. Roy, and K. Chakrabarty. High throughput and volumetric error resilient dilution with digital microfluidic based lab-on-a-chip, 2015. US Patent No. 9,128,014 (Issued on Sept. 08, 2015).

17. Bio-protocol. 2011 (accessed June 3, 2018). `http://www.bio-protocol.org/Default.aspx,`.

18. BioTek. *Rapid Critical Micelle Concentration (CMC) Determination Using Fluorescence Polarization*, 1968 (accessed June 3, 2018). `https://www.biotek.com`.

19. G. M. Bruin. Recent developments in electrokinetically driven analysis on microfabricated devices. *Electrophoresis*, 21(18):3931–3951, 2000.

20. K. Catterall, D. Robertson, S. Hudson, P. R. Teasdale, D. T. Welsh, and R. John. A sensitive, rapid ferricyanide-mediated toxicity bioassay developed using escherichia coli. *Talanta*, 82(2):751–757, 2010.

21. K. Chakrabarty. Design automation and test solutions for digital microfluidic biochips. *IEEE TCAS-I*, 57(1):4–17, 2010.

22. K. Chakrabarty, R. B. Fair, and J. Zeng. Design tools for digital microfluidic biochips: Toward functional diversification and more than moore. *IEEE Trans. on CAD*, 29(7):1001–1017, 2010.

23. K. Chakrabarty and F. Su. *Digital Microfluidic Biochips - Synthesis, Testing, and Reconfiguration Techniques*. CRC Press, 2007.

24. K. Chakrabarty and T. Xu. *Digital Microfluidic Biochips: Design and Optimization*. CRC Press, 2010.

25. T-W. Chiang, C-H. Liu, and J-D. Huang. Graph-based optimal reactant minimization for sample preparation on digital microfluidic biochips. In *Proc. of VLSI-DAT*, pages 1–4, 2013.

26. S. Cho, D-K. Kang, J. Choo, A. Demello, and S-I. Chang. Recent advances in microfluidic technologies for biochemistry and molecular biologys. *BMB Reports*, 44(11):705–712, 2011.

27. S. Cho, H. Moon, and C-J. Kim. Creating, transporting, cutting, and merging liquid droplets by electrowetting-based actuation for digital microfluidic circuits. *Journal of Microelectromechanical Systems*, 12(1):70–80, 2003.

28. N. J. Cira, J. Y. Ho, M. E. Dueck, and D. B. Weibel. A self-loading microfluidic device for determining the minimum inhibitory concentration of antibiotics. *Lab Chip*, 12:1052–1059, 2012.

29. S. W. Dertinger, D. T. Chiu, N. L. Jeon, and G. M. Whitesides. Generation of gradients having complex shapes using microfluidic networks. *Analytical Chemistry*, 73(6):1240–1246, 2001.

30. H. Ding, S. Sadeghi, G. J. Shah, S. Chen, P. Y. Keng, C-J. Kim, and R. M. van Dam. Accurate dispensing of volatile reagents on demand for chemical reactions in ewod chips. *Lab Chip*, 12(18):3331–3340, 2012.

31. T. A. Dinh, S. Yamashita, and T-Y. Ho. A network-flow-based optimal sample preparation algorithm for digital microfluidic biochips. In *Proc. of ASP-DAC*, pages 225–230, 2014.

32. T. A. Dinh, S. Yamashita, T-Y. Ho, and Y. Hara-Azumi. A clique-based approach to find binding and scheduling result in flow-based microfluidic biochips. In *Proc. of ASP-DAC*, pages 199–204, 2013.

33. C. Dong, Y. Jia, J. Gao, T. Chen, P-I. Mak, M-I. Vai, and R. P. Martins. A 3D microblade structure for precise and parallel droplet splitting on digital microfluidic chips. *Lab Chip*, 17:896–904, 2017.

34. R. Drechsler and U. Kühne. Formal modeling and verification of cyber-physical systems. In *Springer Fachmedien Wiesbaden*, 2015.

35. M. Elfar, Z. Zhong, Z. Li, K. Chakrabarty, and M. Pajic. Synthesis of error-recovery protocols for micro-electrode-dot-array digital microfluidic biochips. *ACM Trans. on ECS*, 16(5s):127:1–127:22, 2017.

36. K. S. Elvira, R. Leatherbarrow, J. Edel, and A. deMello. Droplet dispensing in digital microfluidic devices: Assessment of long-term reproducibility. *Biomicrofluidics*, 6(2):022003–022003–10, 2012.

37. M. J. Espy, J. R. Uhl, L. M. Sloan, S. P. Buckwalter, M. F. Jones, E. A. Vetter, J. D. Yao, N. L. Wengenack, J. E. Rosenblatt, F. R. C. 3rd, and T. F. Smith. Real-time pcr in clinical microbiology: Applications for routine laboratory testing. *Clinical Microbiology Reviews*, 19(1):165–256, 2006.

38. R. Evans, L. Luan, N. M. Jokerst, and R. B. Fair. Optical detection heterogeneously integrated with a coplanar digital microfluidic lab-on-a-chip platform. *IEEE Sensors Journal*, pages 423–426, 2007.

39. R. B. Fair. Digital microfluidics: is a true lab-on-a-chip possible? *Microfluidics and Nanofluidics*, 3(3):245–281, 2007.

40. R. B. Fair, V. Srinivasan, H. Ren, P. Paik, V. K. Pamula, and M. G. Pollack. Electrowetting-based on-chip sample processing for integrated microfluidics. In *IEEE International Electron Devices Meeting*, pages 32.5.1–32.5.4, 2003.

41. L. A. Field, B. Deyarmin, C. D. Shriver, D. L. Ellsworth, and R. E. Ellsworth. *Laser Microdissection for Gene Expression Profiling*, pages 17–45. Humana Press, Totowa, NJ, 2011.

42. Y. Fouillet, D. Jary, C. Chabrol, P. Claustre, and C. Peponnet. Digital microfluidic design and optimization of classic and new fluidic functions for lab on a chip systems. *Microfluidics and Nanofluidics*, 4(3):159–165, 2008.

43. P. Gascoyne and J. Vykoukal. Dielectrophoresis-based sample handling in general-purpose programmable diagnostic instruments. *Proc. of IEEE*, 92:22–42, 2004.

44. P. R. Gascoyne and J. V. Vykoukal. Dielectrophoresis-based sample handling in general-purpose programmable diagnostic instruments. *Proc. of IEEE*, 92:22–42, 2004.

45. A. Gefen. *Bioengineering Research of Chronic Wounds: A Multidisciplinary Study Approach*, volume 1. New York, NY, USA: Springer, 2009.

46. J. Gong and C-J. Kim. All-electronic droplet generation on-chip with real-time feedback control for ewod digital microfluidics. *Lab Chip*, 8(6):898–906, 2008.

47. E. J. Griffith, S. Akella, and M. K. Goldberg. Performance characterization of a reconfigurable planar-array digital microfluidic system. *IEEE Trans. on CAD*, 25(2):345–357, 2006.

48. D. Grissom, C. Curtis, and P. Brisk. Interpreting assays with control flow on digital microfluidic biochips. *ACM Journal on ETCS*, 10(3):24:1–24:30, 2014.

49. D. Grissom, C. Curtis, S. Windh, C. Phung, N. Kumar, Z. Zimmerman, K. O'Neal, J. McDaniel, N. Liao, and P. Brisk. An open-source compiler and PCB synthesis tool for digital microfluidic biochips. *Integration*, 51:169–193, 2015.

50. U. Hassan, B. Reddy, G. Damhorst, O. Sonoiki, T. Ghonge, C. Yang, and R. Bashir. A microfluidic biochip for complete blood cell counts at the point-of-care. *Technology*, 3(4):201–213, 2015.

51. K. E. Herold and A. Rasooly. Fabrication and microfluidics. *Lab-on-a-Chip Technology*, 1, 2009.

52. T-Y. Ho. *Design Automation for Digital Microfluidic Biochips: From Fluidic-Level Toward Chip-Level*. CRC Press, 2019.

53. J. W. Hong and S. R. Quake. Integrated nanoliter systems. *Nature Biotechnology*, 21:1179 –1183, 2003.

54. S. Hosic, S. Murthy, and A. Koppes. Microfluidic sample preparation for single cell analysis. *Analytical Chemistry*, 88(1):354–380, 2016.

55. Y-L. Hsieh, T-Y. Ho, and K. Chakrabarty. On-chip biochemical sample preparation using digital microfluidics. In *Proc. of BioCAS*, pages 297–300, 2011.

56. Y-L. Hsieh, T-Y. Ho, and K. Chakrabarty. Design methodology for sample preparation on digital microfluidic biochips. In *Proc. of ICCD*, pages 189–194, 2012.

57. Y-L. Hsieh, T-Y. Ho, and K. Chakrabarty. A reagent-saving mixing algorithm for preparing multiple-target biochemical samples using digital microfluidics. *IEEE Trans. on CAD*, 31(11):1656–1669, 2012.

58. Y-L. Hsieh, T-Y. Ho, and K. Chakrabarty. Biochip synthesis and dynamic error recovery for sample preparation using digital microfluidics. *IEEE Trans. on CAD*, 33(2):183–196, 2014.

59. K. Hu, B-N. Hsu, A. Madison, K. Chakrabarty, and R. B. Fair. Fault detection, real-time error recovery, and experimental demonstration for digital microfluidic biochips. In *Proc. of DATE*, pages 559–564, 2013.

60. J-D. Huang, C-H. Liu, and T-W. Chiang. Reactant minimization during sample preparation on digital microfluidic biochips using skewed mixing trees. In *Proc. of ICCAD*, pages 377–384, 2012.

61. T-W. Huang, T-Y. Ho, and K. Chakrabarty. Reliability-oriented broadcast electrode-addressing for pin-constrained digital microfluidic biochips. In *Proc. of ICCAD*, pages 448–455, 2011.

62. T-W. Huang, C-H. Lin, and T-Y. Ho. A contamination aware droplet routing algorithm for the synthesis of digital microfluidic biochips. *IEEE Trans. on CAD*, 29(11):1682–1695, 2010.

63. H. F. Hull, R. Danila, and K. Ehresmann. Smallpox and bioterrorism: Public-health responses. *The Journal of Laboratory and Clinical Medicine*, 142:221–228, 2003.

64. IBM ILOG CPLEX Optimizer. 1911 (accessed June 3, 2018). `http://www.ibm.com/software/integration/optimization/cplex/`.

65. M. Ibrahim and K. Chakrabarty. Efficient error recovery in cyberphysical digital-microfluidic biochips. *IEEE Trans. on MSCS*, 1:46–58, 2015.

66. M. Ibrahim and K. Chakrabarty. Error recovery in digital microfluidics for personalized medicine. In *Proc. of DATE*, pages 247–252, 2015.

67. M. Ibrahim and K. Chakrabarty. Cyber-physical digital-microfluidic biochips: Bridging the gap between microfluidics and microbiology. *Proc. of IEEE*, pages 1–27, 2017.

68. M. Ibrahim and K. Chakrabarty. Digital-microfluidic biochips for quantitative analysis: Bridging the gap between microfluidics and microbiology. In *Proc. of DATE*, pages 1787–1792, 2017.

69. M. Ibrahim, K. Chakrabarty, and K. Scott. Synthesis of cyberphysical digital-microfluidic biochips for real-time quantitative analysis. *IEEE Trans. on CAD*, 36(5):733–746, 2017.

70. M. Ibrahim, Z. Li, and K. Chakrabarty. *Advances in Design Automation Techniques for Digital-Microfluidic Biochips*, pages 190–223. Springer Fachmedien Wiesbaden, 2015.

71. Epigenie informally informative. 2010 (accessed June 3, 2018). `http://epigenie.com/guide-quick-tips-for-chip-beginners/`.

72. G. I. C Ingram. A note on dilution systems. *Immunology*, 5(4):504–510, 1962.

73. Y-H. Jang, M. J. Hancock, S. B. Kim, S. Selimovic, W. Y. Sim, H. Bae, and A. Khademhosseini. An integrated microfluidic device for two-dimensional combinatorial dilution. *Lab Chip*, 11:3277–3286, 2011.

74. M. J. Jebrail and A. R. Wheeler. Digital microfluidic method for protein extraction by precipitation. *Analytical Chemistry*, 81(1):330–335, 2009.

75. C. Jin, X. Xiong, P. Patra, R. Zhu, and J. Hu. Design and simulation of high-throughput microfluidic droplet dispenser for lab-on-a-chip applications. In *Proc. of COMSOL*, pages 1–7, 2014.

76. S. Jin, H. Jeong, B. Lee, S. Lee, and C. Lee. A programmable microfluidic static droplet array for droplet generation, transportation, fusion, storage, and retrieval. *Lab Chip*, 15:3677–3686, 2015.

77. N. M. Jokerst, L. Luan, S. Palit, M. Royal, S. Dhar, M. Brooke, and T. Tyler. Progress in chip-scale photonic sensing. *IEEE Trans. on BioCAS*, 3:202–211, 2009.

78. O. Keszocze, M. Ibrahim, R. Wille, K. Chakrabarty, and R. Drechsler. Exact synthesis of biomolecular protocols for multiple sample pathways on digital microfluidic biochips. In *Proc. of VLSID*, pages 121–126, 2018.

79. O. Keszocze, Z. Li, A. Grimmer, R. Wille, K. Chakrabarty, and R. Drechsler. Exact routing for micro-electrode-dot-array digital microfluidic biochips. In *Proc. of ASP-DAC*, pages 708–713, 2017.

80. C. Klockenbusch and J. Kast. Optimization of formaldehyde cross-linking for protein interaction analysis of non-tagged integrin $\beta 1$. *Journal of Biomedicine and Biotechnology*, 2010(927585):1–13, 2010.

81. D. Koley and A. J. Bard. Triton x-100 concentration effects on membrane permeability of a single hela cell by scanning electrochemical microscopy (secm). *Proc. of NAS*, 107(39):16783–16787, 2010.

82. A. Kumar, D. Roberts, K. Wood, B. Light, J. Parrillo, S. Sharma, R. Suppes, D. Feinstein, S. Zanotti, L. Taiberg, D. Gurka, A. Kumar, and M. Cheang. Duration of hypotension before initiation of effective antimicrobial therapy is the critical determinant of survival in human septic shock. *Critical Care Medicine*, 34:1589–1596, 2006.

83. S. Kumar, A. Gupta, S. Roy, and B. B. Bhattacharya. Design automation of multiple-demand mixture preparation using a k-array rotary mixer on digital microfluidic biochips. In *Proc. of ICCD*, pages 273–280, 2016.

84. B. Kuswandi, Nuriman, J. Huskens, and W. Verboom. Optical sensing systems for microfluidic devices: A review. *Analytica Chimica Acta*, 601(2):141–155, 2007.

85. K.Y-T. Lai, Y-T. Yang, and C-Y. Lee. An intelligent digital microfluidic processor for biomedical detection. *Journal of Signal Processing Systems*, 78(1):85–93, 2015.

86. C-P. Lee, H-C. Chen, and M-F. Lai. Electrowetting on dielectric driven droplet resonance and mixing enhancement in parallel-plate configuration. *Biomicrofluidics*, 6(1):12814–128148, 2012.

87. K. Lee, C. Kim, B. Ahn, R. Panchapakesan, A. R. Full, L. Nordee, J. Y. Kang, and K. W. Oh. Generalized serial dilution module for monotonic and arbitrary microfluidic gradient generators. *Lab Chip*, 9:709–717, 2009.

88. W. G. Lee, Y-G. Kim, B. G. Chung, U. Demirci, and A. Khademhosseini. Nano/microfluidics for diagnosis of infectious diseases in developing countries. *Advanced Drug Delivery Reviews*, 62(4):449–457, 2010.

89. Y-C. Lei, Y-L. Chen, and J-D. Huang. Reactant cost minimization through target concentration selection on microfluidic biochips. In *Proc. of BioCAS*, pages 58–61, 2016.

90. S. X. Leng, J. E. McElhaney, J. D. Walston, D. Xie, N. S. Fedarko, and G. A. Kuchel. Elisa and multiplex technologies for cytokine measurement in inflammation and aging research. *Journals of Gerontology Series A, Biological Sciences and Medical Sciences*, 63(8):879–884, 2008.

91. A. M. Lesk. *Introduction to Protein Science: Architecture, Function, and Genomics.* Oxford University Press, New York, 2010.

92. Z. Li, K. Y-T. Lai, K. Chakrabarty, T-Y. Ho, and C-Y. Lee. Droplet size-aware and error-correcting sample preparation using micro-electrode-dot-array digital microfluidic biochips. *IEEE Trans. on BioCAS*, 11(6):1380–1391, 2017.

93. Z. Li, K. Y-T. Lai, K. Chakrabarty, T-Y. Ho, and C-Y. Lee. Sample preparation on micro-electrode-dot-array digital microfluidic biochips. In *Proc. of ISVLSI*, pages 146–151, 2017.

94. Z. Li, K. Y-T. Lai, J. McCrone, P-H. Yu, K. Chakrabarty, M. Pajic, T-Y. Ho, and C-Y. Lee. Efficient and adaptive error recovery in a micro-electrode-dot-array digital microfluidic biochip. *IEEE Trans. on CAD*, 37(3):601–614, 2018.

95. Z. Li, K. Y-T. Lai, P-H. Yu, K. Chakrabarty, T-Y. Ho, and C-Y. Lee. Droplet size-aware high-level synthesis for micro-electrode-dot-array digital microfluidic biochips. *IEEE Trans. on BioCAS*, 11(3):612–626, 2017.

96. Z. Li, K. Y-T. Lai, P-H. Yu, K. Chakrabarty, T-Y. Ho, and C-Y. Lee. Structural and functional test methods for micro-electrode-dot-array digital microfluidic biochips. *IEEE Trans. on CAD*, 37(5):968–981, 2018.

97. T-C. Liang, Y-S. Chan, T-Y. Ho, K. Chakrabarty, and C-Y. Lee. Sample preparation for multiple-reactant bioassays on micro-electrode-dot-array biochips. In *Proc. of ASP-DAC*, pages 468–473, 2019.

98. T-C. Liang, Y-S. Chan, T-Y. Ho, K. Chakrabarty, and C-Y. Lee. Multitarget sample preparation using meda biochips. *IEEE Trans. on CAD*, 39(10):2682–2695, 2020.

99. C-H. Liu, H-H. Chang, T-C. Liang, and J-D. Huang. Sample preparation for many-reactant bioassay on DMFBs using common dilution operation sharing. In *Proc. of ICCAD*, pages 615–621, 2013.

100. C-H. Liu, T-W. Chiang, and J-D. Huang. Reactant minimization in sample preparation on digital microfluidic biochips. *IEEE Trans. on CAD*, 34(9):1429–1440, 2015.

101. L. Luan, R. D. Evans, N. M. Jokerst, and R. B. Fair. Integrated optical sensor in a digital microfluidic platform. *IEEE Sensors Journal.*, 8:628–635, 2008.

102. V. N. Luk and A. R. Wheeler. A digital microfluidic approach to proteomic sample processing. *Analytical Chemistry*, 81(11):4524–4530, 2009.

103. M. Lukas. A simulator for digital microfluidic biochips. Master's thesis, Technical University of Denmark, 2010.

104. Y. Luo, B. B. Bhattacharya, T-Y. Ho, and K. Chakrabarty. Design and optimization of a cyberphysical digital-microfluidic biochip for the polymerase chain reaction. *IEEE Trans. On CAD*, 34(1):29–42, 2015.

105. Y. Luo, K. Chakrabarty, and T-Y. Ho. Error recovery in cyberphysical digital microfluidic biochips. *IEEE Trans. on CAD*, 32(1):59–72, 2013.

106. Y. Luo, K. Chakrabarty, and T-Y. Ho. Real-time error recovery in cyberphysical digital-microfluidic biochips using a compact dictionary. *IEEE Trans. on CAD*, 32(12):1839–1852, 2013.

107. Y. Luo, K. Chakrabarty, and T-Y. Ho. Biochemistry synthesis on a cyberphysical digital microfluidics platform under completion-time uncertainties in fluidic operations. *IEEE Trans. on CAD*, 33(6):903–916, 2014.

108. Markets and Markets. *In Vitro Diagnostics/IVD Market By Product (Reagent, Instruments, Software, Service), Technique [Immunoassay, Clinical Chemistry (Lipids, Renal, Thyroid), PCR], Application (Hematology, Diabetes, Nephrology, Oncology), End User (Hospital, POC) – Global Forecast to 2024*, 2019.

109. C. A. Mein, B. J. Barratt, M. G. Dunn, T. Siegmund, A. N. Smith, L. Esposito, S. Nutland, H. E. Stevens, A. J. Wilson, M. S. Phillips, N. Jarvis, S. Law, M. D. Arruda, and J. A. Todd. Evaluation of single nucleotide polymorphism typing with invader on pcr amplicons and its automation. *Genome Research*, 10(3):330–343, 2000.

110. T. Mikeska and A. Dobrovic. Validation of a primer optimisation matrix to improve the performance of reverse transcription – quantitative real-time pcr assays. *BMC Research Notes*, 2(1):112, 2009.

111. D. Mitra, S. Ghoshal, H. Rahaman, K. Chakrabarty, and B. B. Bhattacharya. Test planning in digital microfluidic biochips using efficient eulerization techniques. *Journal of Electronic Testing*, 27(5):657–671, 2011.

112. D. Mitra, S. Ghoshal, H. Rahaman, K. Chakrabarty, and B. B. Bhattacharya. On-line error detection in digital microfluidic biochips. In *Proc. of ATS*, pages 332–337, 2012.

113. D. Mitra, S. Roy, S. Bhattacharjee, K. Chakrabarty, and B. B. Bhattacharya. On-chip sample preparation for multiple targets using digital microfluidics. *IEEE Trans. on CAD*, 33(8):1131–1144, 2014.

114. S. Mondal, E. Hegarty, C. Martin, S. K. Gökçe, N. Ghorashian, and Adela Ben-Yakar. Large-scale microfluidics providing high-resolution and high-throughput screening of caenorhabditis elegans poly-glutamine aggregation model. *Nature Communications*, 7(13023), 2016.

115. F. Mugele and J-C. Baret. Electrowetting: from basics to applications. *Journal of Physics: Condensed Matter*, 17(28):705–774, 2005.

116. J. L. Osborn, B. Lutz, E. Fu, P. Kauffman, D. Y. Stevens, and P. Yager. Microfluidics without pumps: Reinventing the T-sensor and H-filter in paper networks. *Lab Chip*, 10(20):2659–2665, 2010.

117. E. Ouellet, C. Lausted, T. Lin, C. Yang, L. Hood, and E. T. Lagally. Parallel microfluidic surface plasmon resonance imaging arrays. *Lab Chip*, 10:581–588, 2010.

118. N. Pires, T. Dong, U. Hanke, and N. Hoivik. Recent developments in optical detection technologies in lab-on-a-chip devices for biosensing applications. *IEEE Sensors Journal.*, 14(8):15458–15479, 2014.

119. S. Poddar, T. Banerjee, R. Wille, and B. B. Bhattacharya. Robust multi-target sample preparation on meda-biochips obviating waste production. *ACM Trans. on DAES*, 26(1):7:1–7:29, 2020. doi: 10.1145/3414061.

120. S. Poddar, S. Bhattacharjee, S-Y. Fang, T-Y. Ho, and B. B. Bhattacharya. Demand-driven multi-target sample preparation on resource-constrained digital microfluidic biochips. *ACM Trans. on DAES*, 27(1):7:1–7:21, 2021.

121. S. Poddar, S. Bhattarcharjee, S. C. Nandy, K. Chakrabarty, and B. B. Bhattacharya. Optimization of multi-target sample preparation on-demand with digital microfluidic biochips. *IEEE Trans. on CAD*, 38(2):253–266, 2019. doi: 10.1109/TCAD.2018.2808234.

122. S. Poddar, S. Ghoshal, K. Chakrabarty, and B. B. Bhattacharya. Error-correcting sample preparation with cyberphysical digital microfluidic lab-on-chip. *ACM Trans. on DAES*, 22(1):2:1–2:29, 2016. doi: 10.1145/2898999.

123. S. Poddar, R. Wille, H. Rahaman, and B. B. Bhattacharya. Dilution with digital microfluidic biochips: How unbalanced splits corrupt target-concentration. *CoRR*, abs/1901.00353, 2019.

124. S. Poddar, R. Wille, H. Rahaman, and B. B. Bhattacharya. Error-oblivious sample preparation with digital microfluidic lab-on-chip. *IEEE Trans. on CAD*, 38(10):1886–1899, 2019. doi: 10.1109/TCAD.2018.2864263.

125. M. G. Pollack, R. B. Fair, and A. D. Shenderov. Electrowetting-based actuation of liquid droplets for microfluidic applications. *Applied Physics Letters*, 77(11):1725–1726, 2000.

126. M. G. Pollack, A. D. Shenderov, and R. B. Fair. Electrowetting-based actuation of droplets for integrated microfluidics. *Lab Chip*, 2:96–101, 2002.

127. P. Pop, I. E. Araci, and K. Chakrabarty. Continuous-flow biochips: Technology, physical-design methods, and testing. *IEEE Design & Test*, 32(6):8–19, 2015.

128. OpenWetWare Protocols. 2008 (accessed June 3, 2018). https://openwetware.org/wiki/Protocols.

129. H. Ren, V. Srinivasan, and R. B. Fair. Design and testing of an interpolating mixing architecture for electrowetting-based droplet-on-chip chemical dilution. In *Proc. of TRANSDUCERS*, volume 1, pages 619–622, 2003.

130. U. Resch-Genger, M. Grabolle, S. Cavaliere-Jaricot, R. Nitschke, and T. Nann. Quantum dots versus organic dyes as fluorescent labels. *Nature Methods*, 5:763–775, 2008.

131. S. Roy, B. B. Bhattacharya, and K. Chakrabarty. Optimization of dilution and mixing of biochemical samples using digital microfluidic biochips. *IEEE Trans. on CAD*, 29:1696–1708, 2010.

132. S. Roy, B. B. Bhattacharya, and K. Chakrabarty. Waste-aware dilution and mixing of biochemical samples with digital microfluidic biochips. In *Proc. of DATE*, pages 1059–1064, 2011.

133. S. Roy, B. B. Bhattacharya, S. Ghoshal, and K. Chakrabarty. High-throughput dilution engine for sample preparation on digital microfluidic biochips. *IET-CDT*, 8(4):163–171, 2014.

134. S. Roy, B. B. Bhattacharya, S. Ghoshal, and K. Chakrabarty. Theory and analysis of generalized mixing and dilution of biochemical fluids using digital microfluidic biochips. *ACM Journal on ETCS*, 11(1):2:1–2:33, 2014.

135. S. Roy, P. P. Chakrabarti, K. Chakrabarty, and B. B. Bhattacharya. Waste-aware single-target dilution of a biochemical fluid using digital microfluidic biochips. *Integration*, 51:194 – 207, 2015.

136. S. Roy, P. P. Chakrabarti, S. Kumar, K. Chakrabarty, and B. B. Bhattacharya. Layout-aware mixture preparation of biochemical fluids on application-specific digital microfluidic biochips. *ACM Trans. on DAES*, 20(3):45, 2015.

137. A. Saadatpour, S. Lai, G. Guo, and G-C. Yuan. Single-cell analysis in cancer genomics. *Analytical Chemistry*, 31(10):576–586, 2015.

138. E. K. Sackmann, A. L. Fulton, and D. J. Beebe. The present and future role of microfluidics in biomedical research. *Nature*, 507:181–189, 2014.

139. T. H. Schulte, R. L. Bardell, and B. H. Weigl. Microfluidic technologies in clinical diagnostics. *Clinica Chimica Acta*, 321(1):1–10, 2002.

140. L. Shao, W. Li, T-Y. Ho, S. Roy, and H. Yao. Lookup table-based fast reliability-aware sample preparation using digital microfluidic biochips. *IEEE Trans. on CAD*, 39(10):2708–2721, 2020.

141. J-U. Shim, L.F. Olguin, G. Whyte, D. Scott, A. Babtie, C. Abell, W.T. Huck, and F. Hollfelder. Simultaneous determination of gene expression and enzymatic activity in individual bacterial cells in microdroplet compartments. *Journal of the American Chemical Society*, 131(42):15251–6, 2009.

142. Y-J. Shin and J-B. Lee. Machine vision for digital microfluidics. *Review of Scientific Instruments*, 81(1):014302, 2010.

143. R. Sista, Z. Hua, P. Thwar, A. Sudarsan, V. Srinivasan, A. Eckhardt, M. Pollack, and V. K. Pamula. Development of a digital microfluidic platform for point of care testing. *Lab Chip*, 8:2091–2104, 2008.

144. J. M. Smith and A. Heese. Rapid bioassay to measure early reactive oxygen species production in arabidopsis leave tissue in response to living pseudomonas syringae. *Plant Methods*, 10(1):6, 2014.

145. A. W. Solomon et al. A diagnostics platform for the integrated mapping, monitoring, and surveillance of neglected tropical diseases: Rationale and target product profiles. *PLoS Neglected Tropical Diseases*, 6(7):e1746, 2002.

146. H. Song and R. Ismagilov. Millisecond kinetics on a microfluidic chip using nanoliters of reagents. *Journal of the American Chemical Society*, 125(47):14613–14619, 2003.

147. J. H. Song, R. Evans, Y-Y. Lin, B-N. Hsu, and R. B. Fair. A scaling model for electrowetting-on-dielectric microfluidic actuators. *Microfluidics and Nanofluidics*, 7(1):75–89, 2009.

148. M. Spencer, S. Barnes, J. Parada, S. Brown, L. Perri, D. Uettwiller-Geiger, H.B. Johnson, and D. Graham. A primer on on-demand polymerase chain reaction technology. *American Journal of Infection Control*, 43(10):1102–1108, 2015.

149. V. Srinivasan, V. K. Pamula, and R. B. Fair. Droplet-based microfluidic lab-on-a-chip for glucose detection. *Analytica Chimica Acta*, 507(1):145–150, 2004.

150. V. Srinivasan, V. K. Pamula, P. Paik, and R. B. Fair. Protein stamping for maldi mass spectrometry using an electrowetting-based microfluidic platform. *Proc. of SPIE*, 5591:26–32, 2004.

151. V. Srinivasan, V. K. Pamula, M. G. Pollack, and R. B. Fair. Clinical diagnostics on human whole blood, plasma, serum, urine, saliva, sweat, and tears on a digital microfluidic platform. In *Proc. of MicroTAS*, pages 1287–1290, 2003.

152. D. Y. Stevens, C. R. Petri, J. L. Osborn, P. Spicar-Mihalic, K. G. McKenzie, and P. Yager. Enabling a microfluidic immunoassay for the developing world by integration of on-card dry reagent storage. *Lab Chip*, 8(12):2038–2045, 2008.

153. F. Su and K. Chakrabarty. Architectural-level synthesis of digital microfluidics-based biochips. In *Proc. of ICCAD*, pages 223–228, 2004.

154. F. Su and K. Chakrabarty. Unified high-level synthesis and module placement for defect-tolerant microfluidic biochips. In *Proc. of DAC*, pages 825–830, 2005.

155. F. Su and K. Chakrabarty. *Benchmarks for Digital Microfluidic Biochip Design and Synthesis*, 2006. Dept. of Electrical and Computer Engineering, Duke university.

156. F. Su and K. Chakrabarty. Module placement for fault-tolerant microfluidics-based biochips. *ACM Trans. on DAES*, 11(3):682–710, 2006.

157. F. Su, W. L. Hwang, and K. Chakrabarty. Droplet routing in the synthesis of digital microfluidic biochips. In *Proc. of DATE*, pages 323–328, 2006.

158. F. Su, S. Ozev, and K. Chakrabarty. Ensuring the operational health of droplet-based microelectrofluidic biosensor systems. *IEEE Sensors Journal*, 5(4):763–773, 2005.

159. F. Su, S. Ozev, and K. Chakrabarty. Concurrent testing of digital microfluidics-based biochips. *ACM Trans. on DAES*, 11(2):442–464, 2006.

160. S. Sugiura, K. Hattori, and T. Kanamori. Microfluidic serial dilution cell-based assay for analyzing drug dose response over a wide concentration range. *Analytical Chemistry*, 82(19):8278–8282, 2010.

161. K. Sun, Z. Wang, and X. Jiang. Modular microfluidics for gradient generation. *Lab Chip*, 8:1536–1543, 2008.

162. J. R. Taylor. *An Introduction Error Analysis: The Study of Uncertainties in Physical Measurements*. University Science Books, 2007.

163. W. Thies, J. P. Urbanski, T. Thorsen, and S. P. Amarasinghe. Abstraction layers for scalable microfluidic biocomputing. *Natural Computing*, 7(2):255–275, 2008.

164. T-M. Tseng, B. Li, T-Y. Ho, and U. Schlichtmann. Reliability-aware synthesis for flow-based microfluidic biochips by dynamic-device mapping. In *Proc. of DAC*, pages 141:1–141:6, 2015.

165. T-M. Tseng, B. Li, U. Schlichtmann, and T-Y. Ho. Storage and caching: Synthesis of flow-based microfluidic biochips. *IEEE Design & Test*, 32(6):69–75, 2015.

166. T-M. Tseng, M. Li, B. Li, T-Y. Ho, and U. Schlichtmann. Columba: co-layout synthesis for continuous-flow microfluidic biochips. In *Proc. of DAC*, pages 147:1–147:6, 2016.

167. M. Valiadi, S. Kalsi, I. G. Jones, C. Turner, J. M. Sutton, and H. Morgan. Simple and rapid sample preparation system for the molecular detection of antibiotic resistant pathogens in human urine. *Biomedical Microdevices*, 18:18, 2016.

168. Vijay V. Vazirani. *Approximation Algorithms*. Springer, 2010.

169. G. Velve-Casquillas, M. Le Berre, M. Piel, and P. Tran. Microfluidic tools for cell biological research. *Nano Today*, 5(1):28–47, 2010.

170. S. Venkatesh and Z. Memish. Bioterrorism – a new challenge for public health. *International Journal of Antimicrobial Agents*, 21:200–206, 2003.

171. E. Verpoorte and N. F. De Rooij. Microfluidics meets MEMS. *Proc. of IEEE*, 91(6):930–953, 2003.

172. D. Villeneuve, AL. Blankenship, and J. Giesy. Derivation and application of relative potency estimates based on in vitro bioassay results. *Environmental Toxicology and Chemistry*, 19(11):2835–2843, 2000.

173. R. L. Wainwright and M. E. Sexton. A study of sparse matrix representations for solving linear systems in a functional language. *Journal of Functional Programming*, 2(1):61–72, 1992.

174. G. M. Walker, N. Monteiro-Riviere, J. Rouse, and A. T. O'Neill. A linear dilution microfluidic device for cytotoxicity assays. *Lab Chip*, 7:226–232, 2007.

175. S. W. Walker and B. Shapiro. Modeling the fluid dynamics of electrowetting on dielectric (EWOD). *Journal of Microelectromechanical Systems.*, 15(4):986–1000, 2006.

176. G. Wang, C. Ho, W. Hwang, and W. Wang. Droplet manipulations on EWOD microelectrode array architecture, 2014. US Patent 8,834,695.

177. G. Wang, D. Teng, and S. K. Fan. Digital microfluidic operations on micro-electrode dot array architecture. *IET Nanobiotechnology*, 5(4):152–160, 2011.

178. G. Wang, D. Teng, Y-T. Lai, Y-W. Lu, Y. Ho, and C-Y. Lee. Field-programmable lab-on-a-chip based on microelectrode dot array architecture. *IET Nanobiotechnology*, 8(3):163–171, 2014.

179. Q. Wang, Y. Shen, H. Yao, T-Y. Ho, and Y. Cai. Practical functional and washing droplet routing for cross-contamination avoidance in digital microfluidic biochips. In *Proc. of DAC*, pages 1–6, 2014.

180. S. Wang, N. Ji, W. Wang, and Z. Li. Effects of non-ideal fabrication on the dilution performance of serially functioned microfluidic concentration gradient generator. In *Nano/Micro Engineered and Molecular Systems (NEMS)*, pages 169–172, 2010.

181. W. Wang, J. Chen, and J. Zhou. Droplet generating with accurate volume for ewod digital microfluidics. In *Proc. of ASIC*, pages 1–4, 2015.

182. X. Wang, O. Hofmann, R. Das, E-M. Barrett, A-J. deMello, J-C. deMello, and D-D. C. Bradley. Integrated thin-film polymer/fullerene photodetectors for on-chip microfluidic chemiluminescence detection. *Lab Chip*, 7:58–63, 2007.

183. J. Wu, X. Wu, and F. Lin. Recent developments in microfluidics-based chemotaxis studies. *Lab Chip*, 13:2484–2499, 2013.

184. T. Xu, K. Chakrabarty, and V. K. Pamula. Defect-tolerant design and optimization of a digital microfluidic biochip for protein crystallization. *IEEE Trans. on CAD*, 29(4):552–565, 2010.

185. T. Xu, V. K. Pamula, and K. Chakrabarty. Automated, accurate, and inexpensive solution-preparation on a digital microfluidic biochip. In *Proc. of BioCAS*, pages 301–304, 2008.

186. H. Yao, T-Y. Ho, and Y. Cai. PACOR: practical control-layer routing flow with length-matching constraint for flow-based microfluidic biochips. In *Proc. of DAC*, pages 142:1–142:6, 2015.

187. J. Yoshida. *Flash Chemistry: Fast Organic Synthesis in Microsystems*. John Wiley & Sons, Ltd, 2008.

188. Y. Zhao and K. Chakrabarty. Digital microfluidic logic gates and their application to built-in self-test of lab-on-chip. *IEEE Trans. on BioCAS*, 4:250–262, 2010.

189. Y. Zhao and K. Chakrabarty. Cross-contamination avoidance for droplet routing in digital microfluidic biochips. *IEEE Trans. on CAD*, 31(6):817–830, 2012.

190. Y. Zhao and K. Chakrabarty. *Design and Testing of Digital Microfluidic Biochips*. Springer, 2012.

191. Y. Zhao, S. K. Chung, U-C. Yi, and S. K. Cho. Droplet manipulation and microparticle sampling on perforated microfilter membranes. *Journal of Micromechanics and Microengineering*, 18(2):025030, 2008.

192. Y. Zhao, T. Xu, and K. Chakrabarty. Integrated control-path design and error recovery in the synthesis of digital microfluidic lab-on-chip. *ACM Journal on ETCS*, 6(3):11:1–11:28, 2010.

193. Z. Zhong, Z. Li, and K. Chakrabarty. Adaptive and roll-forward error recovery in meda biochips based on droplet-aliquot operations and predictive analysis. *IEEE Trans. on MSCS*, 4(4):577–592, 2018.

194. Z. Zhong, Z. Li, K. Chakrabarty, T-Y. Ho, and C-Y. Lee. Micro-electrode-dot-array digital microfluidic biochips: Technology, design automation, and test techniques. *IEEE Trans. on BioCAS*, 13(2):292–313, 2019.

195. Z. Zhong, R. Wille, and K. Chakrabarty. Robust sample preparation on digital microfluidic biochips. In *Proc. of ASP-DAC*, pages 474–480, 2019.

196. X. Zhou, M. S. Fragala, J. E. McElhaney, and G. A. Kuchel. Conceptual and methodological issues relevant to cytokine and inflammatory marker measurements in clinical research. *Current Opinion in Clinical Nutrition and Metabolic Care*, 13:541–547, 2010.

Index

www.ingramcontent.com/pod-product-compliance
Ingram Content Group UK Ltd.
Pitfield, Milton Keynes, MK11 3LW, UK
UKHW021120180425
457613UK00005B/163

* 9 7 8 1 0 3 2 1 1 3 8 0 7 *